Turkey Culture

Giving the Experience of the Most Successful Turkey Raisers in the United States

by William V. Ross

with an introduction by Jackson Chambers

This work contains material that was originally published in 1901.

This publication is within the Public Domain.

This edition is reprinted for educational purposes
and in accordance with all applicable Federal Laws.

Introduction Copyright 2018 by Jackson Chambers

The World's Largest Selection of Vintage Poultry Books

www.VintagePoultry.com

Self Reliance Books

Get more historic titles on animal and stock breeding, gardening and old fashioned skills by visiting us at:

http://selfreliancebooks.blogspot.com/

Introduction

I am pleased to present yet another title on Poultry.

The work is in the Public Domain and is re-printed here in accordance with Federal Laws.

As with all reprinted books of this age that are intended to perfectly reproduce the original edition, considerable pains and effort had to be undertaken to correct fading and sometimes outright damage to existing proofs of this title. At times, this task is quite monumental, requiring an almost total "rebuilding" of some pages from digital proofs of multiple copies. Despite this, imperfections still sometimes exist in the final proof and may detract from the visual appearance of the text.

I hope you enjoy reading this book as much as I enjoyed making it available to readers again.

Jackson Chambers

INDEX.

Attention, Constant, Necessary for Success	62
Blood, The Best the Most Profitable	61
Breed, The Best	60
Breeders Should Exhibit	55
Breeders, Selecting	69
Shape for	72
Enclose the	23
Coloring, Size, and Weight	88
Cooking Turkeys, Method of	36
Culls, Fewer Among Turkeys than Among Chickens	70
Selling	66
Decline in Turkey Breeding in Eastern States, Reasons for	7
Deformed Birds, Never Breed from	72
Diarrhea and Lice	31
Disease, Symptoms of	52
A Mysterious	59–67
Isolation of Necessary	42
Eggs, Experiment with a Large	76
Hatches from Small	75
Ninety a Season	62
Satisfactory Experiment with Small	75
Small May Hatch Big Turkeys	57
Number to a Clutch	76
Soft-Shelled	78
Exercise, Evils from Lack of	31
Exhibition, Preparing for	68
Express Companies' Rates	37
Feeding, Fall	11–17
And Care	79
Green Food for Growth	25
Management of	12
Turkeys Require Frequent	46
Fighting Toms, Preventive for	59
Foods, Bone and Muscle-Producing	40
For Young Turkeys	14
Proportion of Heat-Producing to Flesh-Forming	40
Golden Rule Should Guide the Fancier	74
Grit, Supply	24
Hatching-Time	22
Housing Turkeys	18–45
Lack of Success, Frequent Causes of	42
Legs, Pink	71
Red-Legged Pullet	73
Lice, Diarrhea and	31
On Turkeys	83
Preventing	43
Sulphur for Injurious	56
Watch Closely for	25
Management	84
Market for Turkeys, Good	53
Marking the Poults	32
Mating	6
Best Method of	63–73

Nesting-Time, At	48
Overfeeding Causes Death	28
Don't Overfeed	24
Pepper, Use of Red	53
Post-Mortem Examinations	63
Poultry Raising, Profits in	61
Prices, Low, Mean Poor Stock	58
Quarters, Filthy	68
Roup, Causes of	80
Rain, Colds	30
Scoring Birds	67
Shipping Dressed Turkeys	35
Show Birds, Selecting	54
Sickness, Preventives and Remedies for	37
Size and Markings, Selecting for	71
Some Important Points	51
Stock, Contaminated Should be Excluded from Markets	36
Breeding from Weakened	28
Judging the	33
Selecting the Breeding	69
Swelled Head and its Remedies	68
Swelling on Hock	79
Swollen Foot on Tom	77
Tapeworms, How Contracted a Question	81
Prevalence of in Turkeys	81
Remedies for	81
Turkeys, Always a Welcome Guest at the Feast	69
Setting on the Ground	74
The Bronze	43
Soon Learn to Know their Home	47
Drooping	77
In Alfalfa Patch	84
Range for	84
Varieties of	5–19
Shipping Dressed	35
Method of Cooking	36
Free-Range the Hardiest	86
To Have Big	87
Bourbon Red	87
Success with	52
The Wild	89
Raise More	25
Late-Hatched	29
Habits of the Wild	90
Feeding the Growing	39
Turkey Raising on the Farm	48
Essentials in	32
Not to be Learned in a Day or a Year	27
Not More Confining than other Occupations	26
Unsuccessful Turkey Breeding	41
Varieties, Characteristics of	33
Popular	5–19–45
Water, Fresh, and Clean Fountains	38
Weight and Size, Distinction Between	63
Must Not be Sacrificed to Fancy Markings	65
Yarding Old Birds	21–29
Fifty Enough in One Yard	26
Young Turkeys, Food for	14
Caring for	22–49
Cooked Food for	56
Feeding	22–39–50
Marking	32
Poults Easier to Raise than Chicks	26
Raising	46–85

BRONZE TURKEYS.

PRACTICAL TURKEY RAISING.

THE RECOGNIZED VARIETIES OF TURKEYS—GENERAL-PURPOSE TURKEYS—MATING—SELECTING BREEDERS—BORROWING AND BUYING POOR MALE BIRDS—ONE MAIN REASON FOR THE DECLINE IN TURKEY BREEDING IN THE EASTERN STATES—FOOD FOR THE YOUNG TURKEYS—FALL FEEDING—MANAGEMENT OF FEEDING—A MISTAKEN IDEA.

BY J. F. CRANGLE.

THERE are six recognized varieties of turkeys in the American Standard of Perfection. For general purposes the Bronze Turkey is considered to be the best, for many reasons. They are very hardy, we might say the hardiest of all turkeys; good layers, and the best of mothers. A Bronze Turkey will lay from eighteen to forty eggs a year under proper management. In breeding turkeys, a person should select stock for the season not later than the first of December preceding, by selecting out the finest shaped birds. One of the most particular points is to be sure that their breast-bones are all straight, as many turkeys have crooked breast-bones. When speaking of the breast-bone, we might say that it is the bone which runs between the legs. A person can easily ascertain whether the bone is crooked or not by catching the turkey and examining the breast-bone by feeling of it. If the stock is selected early in the season, say not later than December, we are quite sure, from experience, better results can be attained. It takes a Bronze Turkey several weeks to get wonted to a place.

It is generally considered that one male bird should be mated to from five to eight hens, but a good, vigorous male is capable of taking care of twenty-five hens, if necessary. But right here we might say that it is not a good plan to mate one male turkey with too many hens, for the simple reason that a male only has one connection with the female. If the male bird, from any cause whatever, should not fertilize the litter of eggs, the best part of the whole season is lost, because if the female has been served by the male bird she will

BRONZE TURKEYS.

go off and make her nest, lay her litter of eggs, and the eggs not being fertilized, the same is lost. So it is not wise to depend too much on one male bird with too many females. One of the best ways I know of to overcome that difficulty is to have two male birds, with any number of hens from five to twenty-five, and allowing one male bird to run with the females every day alternately, a great deal of the risk spoken of above will be avoided; but under no consideration allow both toms to run with the hens at the same time, for as a rule you will generally make a failure. After the first of March you will find that the females will begin to wander a little way from home, and this is about the time they are beginning to look up a place to nest. This will generally occur about two weeks before they begin to lay. If you wish them to lay near by, they can be made to do so, in many cases, by supplying nests for them made out of old brush or boards placed together alongside a stone wall. The females like to hide themselves away when they lay, but with ordinary farmers who only have from five to ten turkeys, it is a very easy matter to find their nests by keeping in a secluded spot and watching the hen. Do not let the female see you, for sometimes she will not go to her nest for hours.

In many of the Eastern States where a few years ago they raised hundreds of tons of turkeys, they now have to import them to meet the demand, as it seems almost impossible to raise them. One of the main causes for this decline in the raising of turkeys is, I think, without a doubt, in-breeding. Farmers, as a rule, do not like to invest a few dollars for a male bird, and they usually go to Tom Jones, or one of their neighbors, and borrow a tom, and this thing has

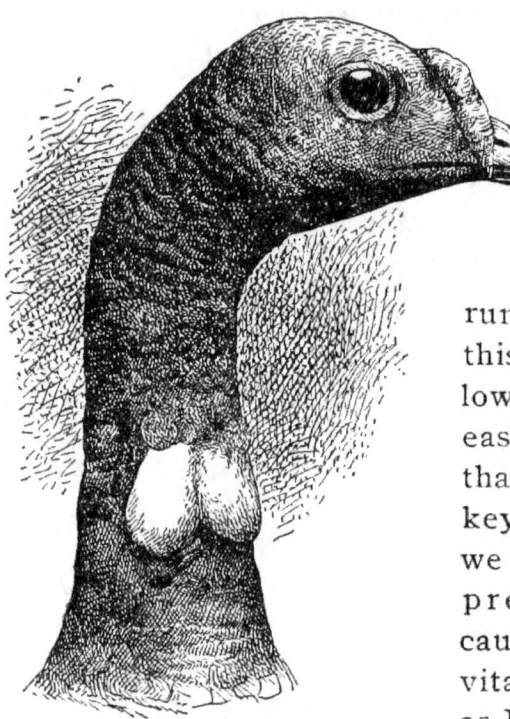

HEAD OF TURKEY.

been done for so many years in the past that the vitality of the turkey has about run out, and by doing this, it has gotten so low that it created disease, and I really think that many of the turkey diseases with which we have to contend at present have been caused by lowering the vitality of the turkey, as I believe there is no other variety of birds in which the vital forces decrease so rapidly by in-breeding as the turkey. I think it is quite possible, under proper management, to raise turkeys in every State in the Union, and I believe that if the farmers in general will be more careful about not in-breeding and will spend a few dollars in order to get a good male bird, and thus introduce new and hardy blood into their stock, they will be able to raise turkeys. I would advise people who have plenty of range for their birds to buy a half-wild gobbler. It is almost impossible to get a pure wild gobbler in this country unless you happen to run across one by accident. About all the people who claim to have wild turkeys have nothing more than half-breeds; but with these you get enough wild blood to make the offspring very much stronger, and this will be very noticeable the first season.

After the hen-turkey commences laying, probably in some sections of the country it will be very cold at night, almost to the freezing-point, and, therefore, under such conditions, the eggs should be gathered every evening and marked with date of getting; then they should be placed in a pan filled with bran with the little end of the egg down, and then the whole placed in a cellar or any cool place, and for two or three days afterward they should be turned every day. It would well to keep them ten or twelve days, but would not as advise keeping them after that time, as perhaps not so many of them would hatch. When these eggs are taken out of the nest, a glass egg should be placed in the nest in order to keep the turkey there, otherwise when she comes back to lay and finding no eggs in the nest, as a rule she will desert the nest and look up another one and lay elsewhere. You will find that a good mother will cover her eggs all up with grass, so that in looking for the nest it is necessary to be very careful, as otherwise you may step into it. An ordinary turkey-hen will cover eighteen eggs. If she lays any more than that number, I take the extra ones and put them under a common hen, setting this hen and the turkey at the same time. If you are afraid your eggs are going to be too old, put the oldest of them under the common hen. She may hatch out a week ahead of the turkey, but as soon as the turkey-hen hatches out her eggs, give all of the poults to her. Be sure there are no lice or vermin on the hen when she hatches. It is also well, when you transfer the poults from the common hen to the turkey-mother, to dust them well with insect powder of some kind. It takes twenty-nine days to hatch turkey eggs. After the twenty-nine days are up, if the turkey does not leave her nest, do not disturb her, because many times she stays on the nest twenty-four hours after the young poults are hatched. The main reason for this is to give the

BRONZE TURKEY.

poults time to gain strength in their legs after hatching. One of the best ways I know of to overcome this is to take a little stale bread moistened with milk, put it gently near the nest, near enough that the turkey hen may get something to eat, and if the young poults are hungry, she will call them out. If she does not appear inclined to move, do not disturb her. It is a very easy matter to find out whether or not she has hatched any poults, for, as a rule, you will find the broken egg-shells scattered near the nest. At the expiration of thirty days if you see no signs of the young turkeys, it is well to investigate the matter by raising the turkey off the nest and ascertaining whether the eggs are unfertile or not. If they should prove to be unfertile, the best way is to shut up the female turkey for four or five days in a coop large enough for her to get a little exercise in, give her food and water and a place to dust herself in. If this happens early in the season, within three or four weeks she will commence to lay again.

One of the best places in which to let the hen-turkey run with the poults is a field where the grass is short. As a rule a pasture is very good—woodland is very suitable. Keep them out of meadows and grain fields until after the grain and hay are harvested, because the wet vegetation is very bad for the young poults, as it chills and sets them back in their growth.

You will always find the finest and strongest turkeys where they are given free range, as much range as possible. As a rule they will not wander far from home. Under proper management you can place turkeys anywhere on the farm you wish, and by teaching them to roost in one particular place, they will come to regard this as their home and will know no other, and you will always find them wherever they have been taught to roost. This can be done by watching them at night for a few times in succession and driving them to the place where you want them to stay, and just before dark they will go up in the trees or on a roost that has been put up.

With the right kind of breeding stock, turkeys at Thanksgiving time should weigh about as follows: Toms, sixteen to twenty pounds; hens, ten to fourteen pounds.

There are two things which have to be done in order to have good success in breeding. One of them is that you have to get the right kind of breeding stock, and the other is to feed them properly. Those are the two main things and the only requirements.

The proper way to feed breeding stock is to be very careful and not over-fatten them. Of course all turkeys are fed principally on corn before Thanksgiving and Christmas, as many of them are dressed for table purposes at that time. As soon as the breeding stock

A FAMILY OF TURKEYS.

has been selected, they should be fed on entirely different lines. The principal food from that time up until the hen commences to lay should be oats. The best way to feed oats is to scald them, but if a person does not care to do this, they can be fed on just ordinary oats with the hulls on. During the very coldest weather in January and February I, perhaps, would feed them a little corn at night, but never any in the morning, and at night-time feed them no more of it than they will eat up cleanly within five or ten minutes after giving it to them. Where the turkeys have range around a barnyard, a person must be very careful not to over-fatten them, and, as a rule, it is only necessary to feed them at night, and under such conditions I would feed corn only about three times a week, all other feeds to be oats.

Another food which we think is necessary for turkeys to keep them in good health is ground charcoal, but be sure not to have it ground fine, as turkeys will eat it better when it is very coarse. On a farm they can ordinarily find all the grit that is necessary for them to have, but I think the eggs will hatch better if the turkeys have oyster shells. These also need to be ground coarsely. If the turkey has not enough lime to properly supply the egg, the shell will be very porous, and many of the germs will die on this account The oyster shell should be set around in small boxes where the turkeys can get at it handily.

Under no consideration breed from a diseased turkey. A turkey that you know has been sick, or is sick, it is much better to kill than have it running with the breeding stock.

When the gobblers mate with the hens, should they be extra heavy and the females extra light, it is well to see that the male does not tear the female on her back. A very good thing to prevent this is to file down the toenails of the male, as many times they will slip off the wings and rip the female open. Should you ever find a female that has been torn open, for it is very easy to discover it by her actions after being tread, or more especially the next day, as she will be lame and her wings will droop, it is best to catch her at once and examine the wound, as most generally they can be saved by sewing up the same. This is not a very difficult matter to do. One person should hold the turkey and another one do the sewing. Pull the feathers away from the edges of the wound, and with some warm water moisten them so that they will stay back while you are putting in the stitches. The wound should be washed out thoroughly with witch hazel, or should you have none, with warm water. Do this with a soft sponge. Then take a long, fine needle with silk thread, draw the edges of the skin around the wound together over it so that the parts meet as they should be. Now commence at one end of the wound, gradually draw the edges of the skin together over the wound as you stitch until the aperture is all closed up. Many times I have taken as many as fifty stitches in one wound. Bathe the wound every two or three days with witch hazel. It is well to keep this particular hen in a pen by herself for two or three days. The period of confinement depends entirely on the size and nature of the wound, but as a rule after two or three days she can be liberated with the rest of the flock.

As a rule many young turkeys are killed by overfeeding. On large farms it is not necessary to feed more than once a day where the turkeys have plenty

FEEDING TIME.

of range. Young turkeys can live on insects and many little grasses which they relish. During the berry season, especially when wild strawberries are ripe, it is a pleasure to watch the little turkeys pick and eat them. In seasons when there is a good crop of grasshoppers the turkeys will live almost entirely on them. Where young turkeys have to be fed, the best food I know of is stale bread, but be sure the bread is not sour. When speaking of stale bread I mean any kind of bread three or four days old. It is very nice to moisten this bread with sweet milk, clabbered milk is also very good for the young turkeys. Put it in a pan on the ground where they can get at it easily. During the very warmest weather of summer it is important to keep the turkeys hungry, for if you do not there is great danger of their having bowel trouble.

The foregoing is a plan for feeding young turkeys where they have free range, but if you are on limited range, the best plan I know is to take three boards, make a triangular pen fourteen inches high, put the young poults inside of it with the mother, and feed them there until they are old enough to jump over the boards. As a rule after that time it will be all right to give them free range.

Many people think it is necessary to put the hen-turkey in a coop to keep her near her young, but this is not true, as she will stay by the little ones and it is almost impossible to drive her away. Where the young poults are confined in a pen as above described, it is necessary, when they are young, to feed them four times a day with stale bread moistened with milk, and give them fresh water or clabbered milk also four times a day. It is also very good to give them a little red

OLD AND YOUNG TURKEYS.

pepper mixed in with the bread about twice a week, as it seems to tone them up a little. Should you find that the young turkeys are drooping and do not seem to pick up, the very first thing to look for is lice. If young turkeys are lousy it is impossible to raise them. Hen-turkeys generally keep free from lice, but some are the same as human beings, very filthy, and all the young poults they will ever have will be lousy, the same as the mother. In looking for lice it is best to examine the little wings, as generally you will find them in the quills, or where the quills should be. They can be killed by using Delmation Powder or any other good insect powder, by dusting them with it. After dusting them with Delmation Powder they will look like a little yellow ball, and in two or three days will commence to brighten up. I feed them stale bread moistened with milk for about four weeks, or until the young poults are able to take hard food; then, when they have limited range and have to be fed, I would feed cracked corn and wheat alternately, but both grains should be scalded and not fed to them until it has cooled. If your poults should have diarrhea from any cause, usually one feed of boiled rice will stop all bowel trouble. Just give them what they will eat up cleanly.

I commence to feed all turkeys the first of October, or not later than the middle of October, in getting them ready for Thanksgiving or for the markets. They should be fed morning and night, but never feed them more than they will eat up cleanly. In fact, a great deal of trouble is caused sometimes by giving them too much to eat. At the time you start to feed them you should only give them sparingly for the first week or ten days, giving it to them night and morning. The principal food from October until the first of December will be corn, either whole or cracked. Above all things do not feed any new corn. I have seen very bad effects result from feeding new corn, and have seen large flocks of turkeys knocked out from this cause. It gives them diarrhea and they get it into their systems and it is very difficult to stop it. Even after you have stopped it you have lost so much time in getting rid of it that they do not recover for weeks.

As a rule most of the turkeys will be fat and fit to kill at Thanksgiving time, but there may be a few that will have to be carried over until Christmas. Many people advocate shutting up turkeys, confining them in order to fatten them. I have tried this plan and found it to be a failure. Many times people can not understand why their turkeys are not as fat as they should be, but almost invariably you will find that they feed them in confinement, and the turkeys could not stand it, especially for a long period of several weeks. The best way we know of is to give them free range, for generally you will find that turkeys will not wander far away from the farm buildings in the fall when the weather becomes cool. They will eat their morning's feed and most of the time during the cold winter will lie down in a warm place and sun themselves. If the turkeys are shut up about five days before the time you wish to dress them, I think they will be in the best possible condition.

A great many people think turkeys need to have shelter in the way of a building. That idea is entirely wrong. The best possible way of keeping turkeys in good health is to have them roost away from buildings, perhaps in trees or in a place sheltered from the winds. I have known turkeys to sit out in trees with the thermometer fifteen to twenty degrees below zero and be much healthier than turkeys that were inside a building. The only thing about it is this: A turkey can stand any kind of cold weather if they are roosting in a lee place where the wind does not blow too hard so that they can keep their heads under their wings. I have seen turkeys in trees outside in a snow and rain storm with the wind blowing at the rate of forty miles an hour, and the colder it got the higher up they would want to go, instead of wanting to come down.

NOTES.

The ordinary turkey of a few years ago, as bred by the average farmer, would weigh about as follows: Toms, about twenty pounds; hens, about six pounds. The standard turkey now weighs: Young toms, eight months old, twenty-four pounds; hens, ten pounds. Where an old-fashioned turkey weighed eight pounds, the modern turkey now weighs twenty pounds. One of the largest of the old-fashioned toms weighed twenty to twenty-two pounds, but the standard tom of to-day weighs thirty-five pounds or more.

Understand, as a rule, a hen-turkey will never want to be served but once.

White Holland Turkeys can stand confinement much better than the Bronze, and will endure it almost as well as a hen. They will lay about the same number of eggs as the Bronze.

The Narragansett is a very nice turkey, but no better than the Bronze; in fact, not so good, because they are not so large and not quite so nicely shaped. But the Narragansetts are next to the Bronze in general utility.

There are the Bronze, Narragansett, White Holland, Black, Buff, and Slate Turkeys. The Buff and Slate Turkeys are, for practical purposes, no good, and are only kept as a matter of fancy for their color and are very small and inferior to either of the other varieties. The only turkeys which are fit to breed are the Bronze, Narragansetts, and White Hollands, simply because the others are too small. The Narragansetts are almost extinct, as there are but very few of them bred. The White Hollands are the gentlest of the three varieties by far.

WILD TURKEYS.

Details of Turkey Growing.

The question to be or not to be a successful turkey raiser is now before us. I got my first turkey eggs the 5th of April this year, last year the last day of March, in 1898 the 6th of April, and in 1897 the 8th of April. In this latitude within a few days of April 1 is as soon as we can look for any eggs. I am getting letters of inquiry asking for my way of feeding and caring for young turkeys.

YARDING THE OLD BIRDS.

The first step to success, I have found by actual experience, is to yard my turkeys. Any grass plot fenced in by a three-foot wire netting, with barbed wire above to make a fence four or five feet high, will do. I have found a three-acre orchard fenced makes a turkey paradise. A row of willows around the outside makes an ideal place for nests. The dry leaves make just the kind of nest material, and it is much easier hunting nests in an acre field or three acres than following the old-fashioned way of letting the turkeys roam without restraint over your own and your neighbor's fields, calling down on your unfortunate head blessings galore. And, by the way, these same blessings proved a lasting help to me. As we had a neighbor whom the sight of a turkey would throw into a spasm, and to avoid the wrath to come, I adopted the plan of yarding and civilizing my turkeys. This proves how what may seem to be your undoing may be a blessing in disguise.

The next thing after the turkeys are yarded is to control these flights of fancy to be outside when you want them to be inside. We began with yard No. 1 and cut the flight feathers on the right wing. Going to No. 2 yard we cut their wing feathers on their left wing. That makes the two yards marked, so if one gets out we can easily tell where they belong and drive them back. Now No. 3 yard must be marked differently, and as we only have a picket fence we shingle them the same as people do with unruly boys. I take a shingle or a very thin piece of board, and make four holes just to fit the wings. Across the back pass a stout string of cloth, through one hole and down around the wing, putting it up through the other and tying securely on top. This made fast in this way prevents their raising their wings up to fly. They will turn somersaults and stand on their head for a while, but, like unruly people, they soon learn to accept the inevitable and realize that the human race with their two hands are a wee bit smarter than the lordly turkey.

HATCHING—TIME, AND FEEDING AND CARING FOR YOUNG.

Now when we have got things in ship shape order the question of hatching and care of the poults is the absorbing thought. I set about two chicken hens and one turkey hen, or put turkey eggs with duck eggs in the incubator and give the turkey the poults to raise. I leave the turkey undisturbed until I see that she begins to get nervous about coming off, then I take the little poults in a well-warmed basket to the kitchen stove and take madam turkey to my louse box, that is painted with a lice killer, diluted a little. I shut her in here for an hour or more. I have a large square dry-goods box, with a square door sawed out, with a lath door made to fit in. I make a pen around the coop of foot-wide boards. Instead of a floor to the coop I put in dry straw and as soon as I turn the poults out I put in clean straw every day. I only keep them cooped three or four days if the weather is good and warm; then I turn them out when the dew is off and let them run until between 4 and 5 o'clock, when I hunt them up and drive them in and feed for the night. At first it is quite a bother to find them at night, but after the little ones learn that they get their supper when you shut them up they will come at your call.

After the ground is dry and warm I move the coops every day instead of putting in dry straw. I forgot to say that before I give the little poults to the hen I grease their heads a little and dust them with Lambert's Death to Lice. I go over them once a week for lice until they get so large it is impossible to handle them. This is when they are six weeks or two months old. We then drive them to the field in the morning and go after them, as we do the cows. In a week or two they will come up themselves. I put fresh-laid eggs in cold water and let them boil one-half an hour or more, and chop them up, shell and all. This is their first food, with a little fine grit sifted in. After a day or two I chop dandelion leaves with the eggs for their breakfast, with a little curd made of sour milk for their dinner. Onion tops and egg and curd make their supper. I mix a little fine grit in every morning in their food and as they grow older give grit a little coarser. This feed, with oatmeal and millet seed, makes up their bill of fare until they are driven to the field and told to help themselves to what they like best.

In two years I have lost only three turkeys by being sick when little. I have tried feeding johnny cake and cornmeal in different ways and always lost my turkeys. I think cornmeal in any form is too hearty for little poults. I keep them a little hungry, feeding only three times a day what they will eat up clean in a short time. A great many people do not look for lice between the quill feathers of the wing, and there is just where the lice set up housekeeping.

ENCLOSE THE BREEDERS.

Turkeys have been raised longer than the memory of those now living can trace, and still the business is in its infancy. People are realizing each year that it is one of the most profitable crops of poultry that can be raised on a farm from a market standpoint. From the fanciers' outlook they are making such great strides in size, weight and plumage, and command such high prices, that they are doubly profitable.

I raise many turkeys, and like the business, and I attribute much of my success to keeping my turkeys yarded through the breeding season. People generally think it a great expense to build a fence that will keep turkeys in, though I do not find it so. A three-foot woven-wire fence with barbed wire above to make the fence four or five feet high, will keep them confined if the flight feathers of one wing are cut. I keep the young turkeys yarded in the same yard until they are six weeks or two months old, which makes it convenient to protect the young poults from the wet, and I can look over them once a week for lice, and never have to be hunting up my turkeys.

Turkeys like a large range as they grow older, but while young, one to three acres make plenty of range. You will soon find out when they get dissatisfied with their quarters, as they will crawl out or fly into your garden or yard, showing that they are anxious to start on their foraging expeditions. If the hay and oats are cut so that they can get around without tramping things down, or finding too much to hide in, we turn them out in the field in the morning and bring them in at night until they learn to come in themselves.

SUPPLY GRIT AND DON'T OVERFEED.

I find if young turkeys are properly fed and kept perfectly clean and free from lice they have very few diseases. Exercise they must have, but very young turkeys can have sufficient exercise on an acre or two, and a great many young turkeys can be saved by enjoying this exercise under your control.

I give a little sharp grit in their feed every morning. I use grit and oyster shell, the larger part grit, as turkeys to be healthy must have it. I have lost hundreds of turkeys I know by not having plenty of grit with which to grind their food. If they get a little sharp grit in their food every morning it keeps their grinding apparatus in perfect order. Very young turkeys do not find the grit of their own accord, and as they grow older they are liable to gorge themselves with the grit as soon as they discover its use, thereby clogging their digestive organs, while a small quantity in their food each morning keeps them in excellent condition.

Overfeeding is another cause of loss in young turkeys. I feed only three times a day for the good reason that I could not possibly find time to feed oftener with the large number I raise. I find it sufficient. They take more exercise if fed less; then when they are fed they are hungry. The time between feeding, too, allows the food to digest and gives the digestion a little rest.

WATCH CLOSELY FOR LICE.

Turkeys are making very rapid growth, and I find the lice are making rapid growth, too. When I take the old turkey off the nest I paint a box with lice killer, put her in and leave her for two hours. I do not shut her in an air-tight compartment, only close enough for the lice killer to thoroughly fumigate her feathers. This kills all the lice and nits. I grease the heads of the little turkeys to destroy the large head lice; I also dust them thoroughly with Lambert's Death to Lice and paint their coops with Carbolineum, but with all the precautions I find I must look over them once a week for lice.

GREEN FOOD FOR GROWTH.

I feed more green food than most people do, as I find it has the same effect on turkeys that it has on ducks. It produces a large frame. I chop dandelion leaves for them in the morning, and at night chop up onions, tops and all. I notice there is never a scrap of the green food left when they are through eating. They make rapid growth when fed this way, besides it is a cheap way to feed them. One of the most essential things during July is to keep the turkeys free from lice. There is considerable work again getting them started to run out on the range and come home at night. If you allow them to run at large and stay out at night, they will wander away to neighbors and sometimes go miles from home; but if they are driven home nightly for a week or two they will soon come home of their own accord, and then your work in the turkey yard is nearly over, as they can take care of themselves, only you must watch that they do not forget to come home.

RAISE MORE TURKEYS.

I hope to hear of a large crop of turkeys being raised this year to help supply the demand for meat and to furnish breeding stock, for a great many farmers who do not now raise turkeys. If I can in any way encourage and help along this great and growing industry it will be a great pleasure for me to do so.

POULTS EASIER TO RAISE THAN CHICKS.

The turkey has been basely slandered and has been considered to be about as stupid as a mule. I have never had any experience in trying to teach a mule to keep his hind feet on the ground when his best friend stood behind him, but I have taught turkeys to respect my wishes and stay on one plantation. People ask me if they are not hard to raise. I raise a larger per cent of those hatched than I do of chickens. For the last two years I have raised over 95 per cent of all turkeys hatched.

FIFTY ENOUGH IN ONE YARD.

I have found that 50 turkeys in a yard or field are enough to do well. If you keep more than that together they are apt to pile in together and smother after they are about a month old. When I get a flock of fifty, I start another drove in another field. I set four or five turkey hens and at the same time give to chicken hens as many turkey eggs as I think the turkeys can take care of. If possible, I set an incubator with chicken eggs. When they all hatch I give the turkeys all the poults and the chicken hens all the incubator chicks, and that makes business lively all around, myself included. Women in the poultry business have very little time for social duties, and the pink and spider-web teas that are so much the fad in fashionable society have to be given in the chicken yard. The turkeys and chickens do not ask the color of their teas, so they get their supper on time. (And if you expect to succeed you have to give them their supper on time and attend to all the details of the business on time.)

NOT MORE CONFINING THAN OTHER OCCUPATIONS.

I have my little poults so they will fly over a board a foot high when but a week old. There are more turkeys killed by overfeeding and lice and want of grit than all other things combined. If you do not keep them near the house so that you can run them under cover when a heavy storm comes up, you are liable to lose a large per cent. I find a large shed with a board floor is fine to run them in in case of sudden storms. Of course, you must stay close at home to meet all these emergencies. It is not more confining than other occupations. The merchant, lawyer, doctor, mechanic and farmer have to confine themselves closely to business, and the poultry raiser, whether for fancy or market, must make a business and work on business principles.

NOT TO BE LEARNED IN A DAY OR A YEAR.

I raise from three different flocks of turkeys, ten hens and one tom in each flock. We have from one acre to three acres fenced in with a three-foot wire netting three inches apart, with barbed wire at the top, making the fence five feet high. Turkeys will never try to fly over a barbed wire fence. They will crawl under it and crowd through it if the wires are not close together, but they never try flying over it. If they ever attempt it they are almost sure to run a barb through their foot, and one experience of that kind is generally enough. I have had them caught that way and hang until dead. I use the breeding yards for the young turkeys until they are large enough to drive out on the range, putting fifty in each yard. At six weeks or two months they are driven on to their summer range, driving them home at night until they have learned the trick of coming home to roost. I did not learn in a day or in a year the art of raising nearly all the turkeys hatched, not until I had lost hundreds each year, I acting as pallbearer and chief mourner, and I assure you I filled the position of mourner admirably, weeping copiously over buried hopes, and those hopes were of a well-filled purse.

I hope I shall not have to meet those turkeys in the next world and be held accountable for my unpardonable ignorance, but perhaps by sincerely repenting my past mistakes the sin of ignorance will be forgiven me.

OVERFEEDING CAUSES DEATH.

A lady writes me that her turkeys are dying. Upon inquiring into the symptoms and the way she feeds I am of the opinion that she is killing her turkeys with kindness by overfeeding. She feeds them five or six times a day. A turkey in a state of nature picks up its feed, a bug or grasshopper at a time, and never gorges itself with food, as it is liable to do when we feed the flock. A duck can be fed all it will eat and as often as it will eat, but if you feed a turkey the same way you are sure to have trouble. A turkey is a voracious eater and will eat as often as you feed it. I can only get time to feed my turkeys three times a day, and as they nearly all live and make rapid growth I think that is all that is necessary.

A neighbor told me that her turkeys were dying, and I sent her word to come and get some Mica Crystal Grit and give them, as I knew she was not giving them any grit. I advised her to put a little in the food every morning. She did so and her turkeys are no longer dying. It was the absence of sharp grit that caused them to die.

BREEDING FROM WEAKENED STOCK.

Another lady writes to tell me that her turkeys were out in a heavy rain and caught cold. They now have roup, or something very like it. She lost about 160 last year with the same disease. I think she would do well to dispose of her stock of turkeys, as they are undoubtedly predisposed to roup, either from a weakened constitution, or it has been inherited by breeding from roupy stock. I think an outward application of Mustang Liniment would be the best outward application. I have ceased trying remedies of all kinds, as I believe many of them kill more than they cure. When I find a remedy that is good I use it.

The work for August in the turkey yard is very light as the turkeys are, or should be, out on the range on farms. I only feed them a little in the morning so that they may be induced to run out in search of food, and a little at night to get them to come home. I am unusually late this year about getting my turkeys started out, as oats have ripened later than usual on account of light rains that have kept them growing, and I can not let them into the hay field until the oats are cut, as I do not want them running through the oats. After they have started out, all I have to do is to bring them home at night and keep on the watch for lice. They go through a corn field, and I have noticed the old turkey and young ones stop and wallow in the loose dirt to dust themselves, so I hardly ever have much trouble with lice when they are out on the range.

YARDING TURKEYS.

Several persons have written about the way I yard turkeys, saying the idea of keeping turkeys yarded through the breeding season is something new. One gentleman says he will keep turkeys if he can succeed in keeping them yarded on four acres. That would be ample room for a good flock to breed from, and plenty of room for a flock of young turkeys to run on for six weeks or two months.

I think there will be more turkeys raised in the future, as they are most profitable poultry, and when sold on the market for Thanksgiving Day cost little to raise, besides being a benefit to the farmer in eating insects and weed seeds.

LATE-HATCHED TURKEYS.

The first of July generally ends the turkey egg business. Occasionally turkeys lay a third clutch of eggs after that time, but I never consider them of much value, as they do not hatch well and the young turkeys never grow very large. I remember one exception to that rule. I had a brood of young turkeys come off about the first of August, and a pullet from that flock weighed sixteen pounds on the 10th of December. I took first premium with her at Dixon, Ill., before the weight was raised in the Standard. That was one pound above Standard weight on a pullet ten days over four months old.

RAISING ONE FAMILY, HATCHING ANOTHER.

I have now a peculiar freak. A turkey by some means had gotten out of my breeding yard and stolen her nest, I think on or near the railroad track. When we first saw her she had four turkeys larger than quail. We could do nothing with her, and decided to let her run. I had the misfortune to have nine turkeys killed by the cars, and as she disappeared about that time we supposed she and her little turkeys were among them. One day as I went along the track to find if my turkeys had strayed there again, I found her sitting on ten eggs and still caring for her first flock, hovering them at night. Later we often saw her come near the building for water for her first brood, so we thought it best to keep water near her. She will hatch this week, and I am curious to know how she will manage two families of children.

The railroad runs the entire length of our place. We have it fenced part of the way turkey tight. Our loss this year has decided us to fence the entire length turkey tight. We have always intended to fence it, but this year the turkeys have behaved well. The extremely hot weather, however, and the lack of water in the creeks, have made them rove in different directions, and this proves a turkey has a memory, as they persist in going where they found water last year. We now keep water in the field where we want them to run and have much less trouble.

RAIN, COLDS, ROUP.

I am getting letters from all parts of the country telling of good hatches, although many have lost a large proportion of those hatched. One lady hatched about three hundred, and her turkeys were caught out in a rain, got a good soaking, caught cold, and roup set in. I think, perhaps, if she had given them at once a warm feed with plenty of cayenne or black pepper they would have come out all right. After my turkeys were as big as prairie chickens they got a good many wettings that did not hurt them in the least. My turkeys have been so healthy, every feather just as straight and smooth as could be. The first thing every one says is, "How healthy your turkeys look." The only road to success with turkeys is to keep them healthy. Give them plenty of exercise, commencing to let them run through the middle of the day at three or four days old; keep the lice off, and give a little grit in their food every morning, with good, clean water to drink. Coop at night until they begin to want to roost.

I could not turn my turkeys out on the range this year until they were two months old, and they were so anxious to get out that I had to let them go. The oats were late, it being after the 15th of July that they were cut, and the men had to drive the turkeys out of them to prevent running over them, and one got killed as it was.

DIARRHEA AND LICE.

A lady wrote that she hatched sixty-six little turkeys and had only ten left. Her turkeys had a diarrhea, a thin, yellowish discharge. This might be from lack of grit. She said she greased them once a week for lice. Too much grease will kill turkeys. I only grease their heads a little for the large head-lice, and dust them with Lambert's Death to Lice. Most of the lice will be found between the quill feathers below the vent, and on large turkeys on the thighs.

EVILS FROM LACK OF EXERCISE.

I learned something about exercise for very young turkeys this year. I hatched some under hens quite early; it was wet and cold, and of the two evils I decided I would not turn them out to run through the day, so I kept them cooped a week or more. When I went to feed them I found one that did not seem to have the use of its left side. I thought it had got hurt in some way and would soon be all right. It got no better, and I still kept them cooped, as it was so cold and wet; then another got that way. They would push themselves around with their right foot as they lay on their left side. When the third one was taken sick I decided it was paralysis of the left side, brought on by lack of exercise, and so I turned them out. Those that had been affected died. It was still cold and wet when my other turkeys began to hatch, and I kept the first lot of chicks cooped perhaps five days, when one of them acted in the same way as the early hatched birds. I turned them out to run through the day, and that was the last of it. This convinced me that it was paralysis brought on by lack of exercise. The peculiar part of it was that it was always the left side affected. My turkeys are making rapid growth out on the range. I feed them a little grain when they come up at night, and we have such quantities of apples that I put the small ones in a box and chop them up with the spade and feed them to all the poultry, and they do enjoy the cool juice these hot, dry days, and the apples keep them in such good health and are so much better for them than all green or even all grasshoppers.

Mrs. Charles Jones.

Essentials in Turkey Raising.

A prominent surgeon once said that there were three essentials in the successful practice of surgery. The first was patience, the second was patience, and the third was patience. Verily, this applies to poultry-raising in all its branches, and I think is particularly applicable to turkey culture.

MARKING THE POULTS.

The youngsters are now big enough to mark, and I prefer a dab of red or yellow paint on the wings to anything else. Leg bands are all right if your birds stay at home and you want to distinguish the blooded from the common stock; but they can too easily be removed. Then, too, if you see a flock of birds together, a glance shows if your fowls are along. Just in front of our house stretches what was supposed to be a wheat field of forty acres. Into this one of my turkey hens went the other day with her flock of ten and came home with sixteen. Grain is a failure in this section this year, and has been for three years. Some of our farmers did not cut the wheat, while others cut with the mower. The ants have eaten out the corn crop, so feed will likely be very dear this winter. In the last two years the farming element, or rather the male part of it, have awakened to the fact that had it not been for their chickens and turkeys many of the actual necessities would have been gone without.

JUDGING THE STOCK.

Now as to judging your stock. The weight of a full-grown Bronze tom should be thirty-six pounds, hen twenty, cockerel twenty-five, pullet sixteen. Of course, any deformity, such as crooked feet, beak, breast bone, or wry tail, disqualifies the bird for show. Altogether white or black feathers or the absence of grey bars on wings counts you out. The beautiful bronze sheen should be prominent in male, but does not show so brilliantly in female. The weight of the White Holland is, cock twenty-six, hen sixteen, cockerel sixteen, pullet ten. Besides being snowy white, these birds should have pink legs and feet. Believing these two varieties to be bred the most extensively, I will not give other standard, but any one wishing same can obtain it for the asking and a stamp.

CHARACTERISTICS OF VARIETIES.

Personally, I think if one must allow his birds to roam on others' property that he should breed something different from his neighbors. The Bronze is the heaviest bird, and is what I breed. The White Holland turkeys are said to be non-roamers, but my observation of a neighbor's proves that they wander as much as any other fowl that has to make its own living. To be sure, the blacks are sports from the wild turkey, consequently are very shy, although they are hardy if not kept in confinement. In parts of Ohio, Tennessee and Kentucky a red turkey is bred, from which, I think, has sprung the buff turkey. Let me impress on you not to over-feed

CRESTED TURKEY.

the young stock while the hot weather is on; also see that they have plenty of clean water to drink. This latter will keep them at home when nothing else will.

* * * * * * * *

Now is the season when the turkeys are beginning to stay around the buildings to be fed instead of wandering away and foraging. Begin gradually, then feed them all they will eat, unless the weather be very warm, when there is danger of over-feeding. Should any birds develop bowel complaint, remove all drinking water and substitute a dish containing this drink: One-quarter ounce of copperas to a quart of water. They at first refuse it, but as the fever will be high they will find themselves compelled to drink.

SHIPPING DRESSED TURKEYS.

In shipping dressed poultry there are some essentials to be considered. The prospective city buyer may not know a dry picked fowl from the scalded one, but the commission merchant does. Now, the fowl should be killed by inserting a knife in the roof of the mouth, thereby penetrating the brain. Hang the bird up by the feet to bleed out. When this has stopped, dip hastily in hot water, then at once in cold water and pick. This process hardens the flesh and makes them easy to pick. The packing boxes should be lined with white paper and fowls laid in, alternating head and feet; that is, if ten birds constitute a row, have five heads point one way and five the other. In this way they fit snugly. Some dealers like a ruffle of paper added at the knee. Lay a piece of paper over this layer and proceed to fill up the box, not crowding, but fitting snugly. The best paper to use can be obtained very cheaply at any newspaper office, and is such as is used in printing. This paper is absorbent also. This may sound like a good deal of bother, but will repay you many times over. Then, too, you will always find the commission men want your stock. Boxes are to be preferred to barrels because the goods can be displayed to better advantage. This method of packing holds good in all kinds of poultry.

METHOD OF COOKING TURKEYS.

The turkey is called king of birds, yet how often we find them poorly served. In cooking allow twenty minutes to the pound and thirty minutes' grace. Then if the bird is not three or four years old it will be done to a turn. The addition of hickory-nut kernels to the filling is a great improvement. They cook so soft you cannot tell them from the crumbs. Many cooks spoil a beautifully cooked fowl with a dry filling. Moisten the crumbs with cold water and allow two teaspoons baking powder to every quart of crumbs.

* * * * * * * *

Many times one sees in the poultry papers inquiries as to how much damage a stray bird can do a pen in a short while. We have had an experience along that line that might interest some, so will give it. Our tom met with an accident that incapacitated him for service just as the hens were about to lay the third "clutch" of eggs. I was desirous of some late hatched stock, but decided it was out of the question. The particular hen I wanted to breed from had laid five eggs, when a neighboring farmer came in and told us to come and get his tom. He was brought, and stayed one night, then went home. The result of his visit is eight little poults. Some of the papers are telling just now how a man in the West hires out a flock of a thousand turkeys to catch his neighbors' grasshoppers and is coining money. Maybe it is true, but if it is, his neighbors are different from mine.

CONTAMINATED STOCK SHOULD BE EXCLUDED FROM THE MARKETS.

The poultry organizations are advancing in so many ways that it seems queer to me that no effort is made to exclude contaminated stock from the markets. I have particular reference to roup. We buy a fowl apparently in health, and with wet weather it develops roup. We wrestle with the disgusting disease and may think we have cured it, but that bird's progeny will inherit it and before we realize what has happened our whole flock is polluted. It is claimed by some that roup in fowls is the same as diphtheria in the human system; also that a diseased fowl will produce that disease in people if eaten. I think cold steel the best cure, but if your birds appear to have only a cold, house them tightly and throw crude carbolic acid on live coals, then make your exit as speedily as possible. This will produce a dense smoke and cause the birds to almost sneeze their heads off, but will cure the cold, besides acting as a disinfectant to the house.

EXPRESS COMPANIES' RATES.

Express companies now make a difference between market and breeding fowls insomuch that expressage is quite an item. The birds are supposed to get better care in transportation, but you will note that the particular expressman who waters and feeds them en route asks a small fee. I have never had this happen to me, but we have shipped stock that was fed and watered on the way; then the agent asked 10 cents for his service. I should just refuse point blank to pay it. The company is obliged to hold packages twenty-four hours, and in that time look after them, so there is no occasion for charge, be it ever so small.

* * * * * * * *

PREVENTIVES AND REMEDIES FOR SICKNESS.

At this season of the year the turkey toms are liable to fall sick, so will give some preventives. To begin with, when the first mild days of spring come their corn allowance should be curtailed, as that and the breeding season are almost sure to cause sickness. However, if this has been neglected, we must try and prevent the trouble.

The first symptom you will notice is a white diarrhea and a tendency to sleep. Sometimes they will not come off the roost. Make them come down and give pills of venetian red, held together with raw egg. It will take two people to do this, one to hold the bird and the other to handle the pills. Make them as thick and long as your thumb and give six, morning and night, for three or four days. This, with a little alum or turpentine in the drinking water, will be sufficient if you do not let the disease get a start. If the bird has been ailing a day or two, add a two-grain quinine pill to the other preparation morning and night and feed only wheat for a few days. We have seen a thirty-three-pound tom that had been pulled down to twelve pounds by the discharge, pulled through by this method. He had lost all interest in life, so the lice decided to hold a convention on his poor, bony carcass. Little red fellows, long yellow ones, and the big, hairy old stagers, were there. He was painted with alcohol in which fish berries had been soaked. This not only kills the lice, but all nits and eggs, and can also be used to advantage on a person's head when they happen to become infested. But in using this bear in mind that it is deadly poison and should be handled with care. It leaves no injurious effects and can be used freely, as the alcohol soon evaporates.

The little poults that have lived until six weeks will now have another battle for life. This is the season when the head turns red, and is called "pushing the red."

FRESH WATER AND CLEAN FOUNTAINS.

See that their drinking vessels are kept clean and supplied with fresh water, and above all do not overfeed, but rather keep them hungry until eight or ten weeks of age.

Thinking to introduce some new blood into my flock, I sent to Indiana for eggs. They hatched fairly well, but those little Western turkeys were as wild as quail, and even now, while they run to meet me, they do not want to be touched. My own birds climb right into my lap to see what there is to eat. This wildness annoys me, for there is no reason on earth why chickens and turkeys should not both be tame and docile.

Last Sunday morning one of our hens that has been spoiling for a fight for some time attempted to whip another hen with ten little poults. The fight began by the instigator saying something to the mother turk, and in the twinkling of an eye every little poult disappeared. We put a stop to the fight, but from all indications there is more trouble brewing. I know of no way to stop a belligerent turkey hen, for they carry resentment a long time.

MAUDE VON PLEES.

Feeding the Growing Turkeys.

A writer, whose letter is given below, requests information regarding growing turkeys, and suggests a certain combination of food. It is the various mixtures of foods to which we desire to call attention, for which reason we present his letter, which reads as follows:

"I would like to ask a question in regard to feeding young turkeys; as we have sixty-four now, and will probably have 100 soon. We (my wife and myself) would like to keep them in good growing condition.

"How would one part of oil cake or linseed meal and three parts of wheat bran do? And after what age? The turkeys have plenty of range, and we raised forty-two last year, but did not feed the above."

The writer of this letter gives his reason for wishing to change his method of feeding—he wishes to force the turkeys in growth.

After a young turkey "shoots the red," and is past the danger stage, it becomes a hardy bird. Naturally the turkey seeks its food over a wide area, and in so doing secures a variety. It will accept seeds, tender grass, and all kinds of insects. Even the green worm which is found on the tobacco and tomato plants will be acceptable, while grasshoppers provide a feast. The foods secured by turkeys are both carbonaceous and nitrogenous; it consists also of animal and vegetable matter, with a proportion of mineral constituents.

If the turkey is in a limestone section, or the range provides an abundance, it will procure more food during the day than may be supposed; that is, the crop will either be filled several times during the day, or the food sufficient for filling the crop several times will be digested as fast as eaten. If food is provided by the owner it should be simply to induce the turkeys to come up at night in expectation of the reward.

BONE AND MUSCLE-PRODUCING FOODS.

Our correspondent suggests one part of linseed meal and three parts bran. He has not tested these foods for the purpose, and that is why he deemed it best to write. We understand his motive—he wishes to feed for bone and muscle. As he states, he wants growth, for he can put fat on the turkeys later.

Let us examine the selected foods and their value. Linseed meal contains nearly 6 per cent of mineral elements (bone-making material), or 120 pounds per ton. Also 33 per cent of protein (muscle-producing material) and 39 per cent of fat-forming elements. Bran contains 5 per cent of mineral matter, 16 per cent of protein and 53 per cent of fat and heat producers.

These foods are harmless and will be beneficial to turkeys three months old and over. Give them a full meal at night, but the proportions of linseed meal should be one to ten of bran at first, gradually increasing the proportion of linseed until in six weeks it is one to four. Linseed meal is laxative, and may not prove beneficial if given in very large quantities at the beginning.

Cottonseed meal is not so wholesome, for the reason that the cotton ball is not fully matured when picked and the seed is not as advanced as that of flax at the time of harvesting.

PROPORTION OF HEAT-PRODUCING TO FLESH-FORMING FOODS.

The food may be given at night on clean boards. It is somewhat oily and sticky, but the bran serves to divide it. All linseed products used for food are ready cooked, and the new process linseed meal contains less oil than the old process product.

In making up foods one may give from three to six times as much of the heat producers as of flesh formers, according to circumstances. It should be remembered that one pound of fat is equivalent to two and a half pounds of starch. Mineral matter contains a large proportion of lime. Beans, clover, peas and the gluten meals are rich in protein.

A young turkey does not readily fatten until nearly matured, the food being converted into bone and muscle. It makes the frame first and takes on the muscle and fat afterwards. Bone and muscle-producing foods are consequently excellent for young turkeys.

Unsuccessful Turkey Breeding.

"We have been surprised to find how great a proportion of those who attempt to raise turkeys use small and immature birds for breeders. Many kill their earliest and best birds for the market and keep for breeding those that are too small or too late to be salable. They kill the goose that lays the golden eggs. In buying a new gobbler or a few hens to change the blood they choose late-hatched, immature turkeys because they cost less. The reason sometimes given for this is that old hens are too cunning about stealing their nests and that young turkeys lay earlier. This practice is not confined to the poorest and least intelligent people, as would be expected, but is followed by those well informed and who appreciate and pay for a well-bred horse or cow. If such a course was followed with horses and cattle the best stock in existence would be ruined in a few generations. Many who know that turkeys two years old or older give the strongest and largest young continue to kill off the young hens for market after breeding from them one season. There seems to be a dread of having something too old or unsalable left on their hands. To breed from immature or poor specimens is to violate one of the first laws of breeding. Selection of the best for generations has given us the improved and most profitable breeds of stock. The hereditary influence of such selection is of great value. The most inferior bird out of a flock of such blood may 'throw back' and breed very fine stock and do better than a much finer specimen from a poorly bred strain, but the repeated selection of inferior birds for a number of generations makes this inferiority hereditary.

FREQUENT CAUSES OF LACK OF SUCCESS MAY BE TRACED TO THE PARENT STOCK.

"The future stock depends almost entirely on the parent birds or their ancestry. If valuable birds are used for breeding their offspring will be like them and amply repay the extra expense. The best are none too good and are the cheapest.

"Crandall Brothers, having used Western gobblers furnished by Mr. Vose, raised so many more turkeys in consequence that they estimate the benefit derived the first season at $100. It would have been economy for them to have paid $50 for the two gobblers rather than use the kind of stock they had previously bred from. This expenditure would have paid the first season, to say nothing about their improvement in their breeding stock for the future. Many breeding turkeys are over-fat in the spring—have been over-fed or given too fattening food. Quite frequently they die at this time as the result of over-feeding. The progeny of over-fat birds are less vigorous. Late-hatched hens that are growing all the time need more food; cannot store up a surplus, and lay earlier because they are thin. Feed the old hens clover and less carbonaceous food in the latter part of the winter and they will give better satisfaction. Corn is all right when turkeys can find their own green food and insect ration to go with it, but when they get little exercise and can get nothing else to eat they become abnormally fat.

ISOLATION OF DISEASED BIRDS NECESSARY.

"If a turkey becomes sick and is allowed to roam with the others, and to eat, drink and roost in the same places, the others will probably have that trouble very soon. If a flock becomes diseased, the land which they wander over may become contaminated and infect other flocks that occupy the same ground. Therefore stamp out disease when it first appears. Let every turkey raiser be a board of health; quarantine or kill and bury deep all sick fowls and disinfect what they have contaminated. Prevention of the spread of disease is possible. Doctoring very sick turkeys is rarely practicable. If turkeys are kept where they may drink from stagnant pools in the barnyard, near the pig pen, privy vault, or from the sink drain, sudden and fatal attacks of bowel trouble should be expected among them. A running stream is of great value on a turkey farm. If brine is poured out and they drink it, or they pick up pieces of salt, salt meat or salt fish, death usually follows."

REPORT OF R. I. EXPERIMENT STATION.

The Bronze Turkey.

There is more profit in raising turkeys than any poultry raised on the farm, but occasionally we hear people say, "there was nothing in turkey raising for me," and nine times out of ten the fault might have been traced to the management, kind of stock, etc.

On every hand we see the common or scrub stock used. We believe in thoroughbreds, even in cats. Turkeys are just as easy to raise as chickens, but we must use care and not inbreed, as so many do. Inbreeding is more fatal with turkeys than with hens. Procure new stock each year, either in the shape of a thoroughbred gobbler or eggs from a reliable breeder of thoroughbreds, and you will find your stock improve and be strong, vigorous and mature quickly.

PREVENTING LICE.

Lice will kill young turkeys quicker than anything else. The use of some good insect powder on the old hens before hatching will prove a great help in preventing lice on the poults.

Improper feeding is another cause for delicate turkeys. Corn is usually fed too heavily to the hens dur-

NARRAGANSETT TURKEY.

OSCELLATED TURKEY.

ing the winter and the old turkeys are apt to be very fat when they commence to lay. Toward spring the hens should not be allowed to run to the corn bin, but should be on a regular egg ration, for fat hens and fertile eggs will not go together by any means.

HOUSING THE TURKEYS.

We do not believe in confining the hens in a close run during the laying season. The old turkeys can be confined in a large shady yard if it is not desired to hunt for the eggs. Some turkey breeders claim that turkeys do much better if hatched by a turkey hen, but our experience has been that turkeys raised with chicken hens paid us much better than those raised with turkey hens. True, they seem to grow faster and thrive better if raised with turkey hens; but when raised with chicken hens they practice the habit of coming up for their meals and do not wander off from home and "come up missing" so often. At selling time we receive a greater income from those raised by chicken hens than from those raised by turkey hens.

After the turkey commences to sit, erect a temporary cover over her to protect her from rain and storms. There is no question but turkey raising is very profitable, especially where one is located on a large farm. Turkeys are largely self-supporting, and although with some there is a difficulty in bringing them to the two months age, yet they are very hardy thereafter.

THE POPULAR VARIETY.

The most profitable and by far the most popular turkey is the Mammoth Bronze. It is the largest variety, and will outweigh any other variety of the same age by several pounds. They cost no more to raise, and therefore are most profitable. They are the hardiest and most extensively raised of any breed. They do not attain their full size and weight until about three years old. At maturity the hens weigh from fifteen to twenty pounds and gobblers thirty-five to forty pounds each. They bear confinement to yards remarkably well, and the young are easily raised, if proper care is given them. They are excellent layers and good mothers.

The plumage of the male turkey on back and breast is of a brilliant bronzy hue, which glistens in the sunlight like burnished gold. Wing coverts are a beautiful rich bronze, the feathers terminating in a wide, bronzy band across the wings when folded. The plumage of the female is similar to the male, but not so brilliant. Who would not be proud of a nice flock of Bronze turkeys, regardless of the profit they produce?

J. C. Clipp.

Raising Young Turkeys.

Sometimes the turkey hens will begin laying early if the season is mild. In caring for young turkeys much depends on the feed for the first two months. The first food should be stale bread soaked in milk; also chopped onion-tops, and curd made from soured or clabbered milk by scalding it over the fire. To this add a little black pepper three times a week, and feed four times a day the first month. Hard-boiled eggs may be given three times a week, but do not give them too much.

TURKEYS REQUIRE FREQUENT FEEDING.

Turkeys require feed oftener than young chickens. Give them all the milk they can drink and plenty of fresh water. Give small grains of any kind for a change; millet seed and pinhead oatmeal are excellent the first two weeks. Corn bread mixed with sour milk is a good change. Never feed raw corn meal, as it is not beneficial, and never leave feed to remain, as it may become sour; but give only as they will eat up clean at each meal. A little fresh meat, finely chopped, three times a week, may be allowed. When old enough to eat corn feed anything they will eat, as after that the danger is over in regard to feed. Keep the coops clean and dry. Keep their drinking cups clean, and do not expose turkeys to rain or dew, as they are very tender in regard to dampness; but in fair weather let them have range in the daytime, confining them to their coops at night. Give plenty of sand and sharp gravel. Give them a dust-bath of sifted coal ashes; it will make chicken lice hunt other quarters, and use the lice remedies whenever necessary. Set the turkey eggs under common hens. They make good mothers, as they do not stray far from home, and can be confined with less trouble in small coops, allowing one hen to each coop. If kept confined a few days the hen will take her own coop at night. If the hen discards them very young, as is sometimes the case, drive them to their coops until they can fly to roost. An important point is to examine carefully twice a week for the large lice on the heads, a single one of which will kill a young turkey. One-half of the young turkeys die from this cause. These lice come from the hens. The remedy is one or two drops of melted lard, well rubbed in on the head, but be careful and use but very little, as grease is fatal to both young turkeys and chicks.

SOON LEARN TO KNOWN THEIR HOME.

Turkeys will soon learn where they belong if care is used at first, as they can be taught to remain near the barnyard, and the time to begin with them is when they are young. If the wings of the adults are cut they will not fly over a high fence. They can easily be made to thrive on a large lot. It is an excellent plan to feed them twice a day at one place, so as to have them expect their meals and come up for the food, but the morning meal should consist of only about a gill of wheat. At night give a full meal, one night wheat and the next night chopped meat. Where there are many trees they cannot easily be induced to go under shelter, but if the young turkeys are taught to go up at night, and are not allowed to remain outside, they will always come up; but that would necessitate the removal of the old one after the young turkeys are three months old. Much depends on the forage. Turkeys like grass seeds and insects, and will seek such foods if they do not have them on the ground. They will not bear close confinement, but will thrive on a large piece of ground. It does not pay to allow them to stray off if foxes, dogs or other enemies are numerous.—Farm and Fireside.

Turkey Raising on the Farm.

The other day I chanced upon an article by a well-known turkey breeder, in which, after extolling the merits of his own breed, the Bronze, dealing out a modicum of praise to the Narragansett, mentioning the White Holland with a certain degree of respect, and simply naming the Black, he remarks: "Buff and Slate turkeys, for practical purposes, are 'no good,' and are kept only as a matter of fancy for their color, and are very small and inferior." I have been breeding Slate turkeys for several years, and I know from practical experience that they are good utility birds, as well as being of a color which is unlike my neighbors' turkeys, consequently easy to recognize when the flocks mix in the fall. That they are very beautiful and attractive is certainly not against them. At one time I bred them for size as well as feathering, but found that the young birds were not ready for market, even at Thanksgiving, and some of them were hardly marketable at Christmas. I secured a tom, last year, with great width of back and fullness of breast, and the young stock last fall were fat and plump at Thanksgiving, and in every way desirable. The flock were nearly all hens, and weighed from eight to twelve pounds, full dressed, which, for the retail trade to which I cater, is most acceptable. Even the hotels will pay two or three cents per pound less for turkeys from twenty pounds up than for smaller birds. The strain with broad backs, full breasts, stout shanks and moderate size, which careful breeding will produce, is the best with which I am acquainted for private families and small hotels. Turkeys feel the results of inbreeding more than most kinds of poultry, and new blood should be obtained every year, if possible. Where it is desirable to change once in two years only, it is better to breed the same tom to his daughters, rather than to save a young tom to breed to his sisters.

AT NESTING-TIME.

When the turkeys show signs of nest hunting it is wise to put barrels or roomy boxes in secluded places.

putting plenty of marsh hay or straw, without chaff, in them. Make the nests firm, and put a bar across the front, which will keep the eggs from rolling out. Be sure that the nesting-place will not hold water; an auger hole or two will make this positive. A nest egg or two in the box will help to attract Madame Turkey. I save hens' eggs which I suspect of being addled, boil them hard, mark them "Boiled" in large letters, and use them for nest eggs, as they are more attractive to the hens than glass eggs. Some turkeys can be coaxed into the prepared nests by watching them and not allowing them to wander far from them. Turkeys which have been kept tame are much easier to manage than those which are wild. My own birds will allow me to approach their nests when they are laying, and even to take out their eggs, without running away. When the first turkey to begin laying has laid from fifteen to twenty eggs it is well to begin setting the turkey eggs under good, reliable common hens. Put the nests where they are secure from marauders, and give the hens from seven to nine eggs, according to their size. When the turkey hens become broody give them from seventeen to twenty-one of these partially incubated eggs, and when they hatch give all the poults to the turkeys. The hens and the turkey should be well dusted with insect powder before they are given eggs, and, if possible, two or three times during incubation. This is easy with hens, but it is sometimes difficult to lift a turkey from the nest without endangering her eggs.

CARING FOR THE YOUNG.

When the young turkeys begin to hatch remove them in a warm basket to the house and dust them with insect powder, but do not cover them, for they are very delicate when first hatched, and may smother. If the hen becomes nervous while removing the poults, take her out of the nest, and have an attendant hold her and give her a good dusting with the powder, while you imprison the shy, soft little turkeys. Leave a turkey or a pipped egg in the nest to keep the mother happy, put her at the front of the nest and go away quickly. The best plan I have ever tried for the first two or three days with young turkeys was to take a large dry-goods box, slat it across the front, making a door to admit the turkey. Take off one or two boards at the back and cut the side boards on a slant; put on a roof which projects three or four inches in front and back. Old oilcloth, zinc or sheet-iron which will turn water, answers to finish the roof and make it water-tight. Make a pen in front of the house by setting two boards with one end of each against the side of the house and let the other ends come together at a point; fasten securely with stakes, and make sure that there is no place where the foolish little turkeys can hurt themselves or get fast. Have a twelve-inch board in front of the coop on the ground, fitted so that it may lie close to the coop for the youngsters to run out on, or be fastened up against the slats to prevent their coming out at all. Leather hinges do very well on the door, and a leather strap with a slit cut through to pass over a fence staple gives a secure fastening, if a nail is put through the staple after the leather is slipped on. The board may be secured across the front by two large wooden buttons. When incubation is finished, take the hen turkey from the nest, put her in some inclosure, give her a generous feed of corn and water and leave her for an hour; then put her into the box-house, grease the head of each little poult with soft lard, and when ready take them to their mother, put them into the box as quickly and quietly as possible and leave them to become accustomed to the new home.

FEEDING THE POULTS.

The poults require no food for the first day. For the early feedings I put a little fine grit in the bottom of a shallow jelly-cake pan, and scatter the food over it, bread soaked in cold, sweet milk, and squeezing out dry, being the first feed; then sour milk scalded and the curd drained dry and mixed with the bread crumbs. The food is usually seasoned with black pepper. After the first few days the turkey is allowed to run out during the day, but is fastened into the house at night.

When the poults are a week old the whole brood are given their liberty, but are watched and put under cover in case of storm, and are housed at night. In two or three weeks the hen will insist that she knows more about turkey raising than her caretaker, and will not permit herself to be shut into the house, and when this takes place let her have her own way, but keep watch of her and feed the brood two or three times a day. By this time the poults regard their caretaker as a second mother, and will come at the call. In case of hard, sudden rains, I have carried the whole drenched flock to the kitchen, and warmed and dried them, even when as large as common hens. Turkeys need the tenderest care till they are from six weeks to two months old, when they can find for themselves; but it is well to coax them home every night by generous feeding at a fixed hour. They will not often forget the habit if it is a custom. If they do not come themselves, find them and bring them home.

SOME IMPORTANT POINTS.

The utmost gentleness should be used to all kinds of poultry. It pays to have them tame. Cleanliness is also absolutely necessary. The coops for the birds must be thoroughly cleaned often, and if the floor is kept covered with dry sand, in which a little carbolic acid has been mixed, it is well. It is not wise to have the turkeys and chickens run together, though when the brood of young turkeys seems too small for the hen she will adopt chicks, or even ducks, and give them the same care she does her own young. After the young turkeys begin to wander I mix their sour milk curd with ground grain, cracked corn, boiled wheat screenings and anything which I give the chicks, except I do not give them meat.

Turkeys are a profitable bird, and though the first few weeks of their existence are anxious ones, they are soon able to care for themselves and pick up the **greater** part of their food from the fields and forests.

S. A. LITTLE.

Success With Turkeys.

The cry of cholera among turkeys comes to me from many persons, and these are not confined to a given locality or State. From Mississippi, Wisconsin, Pennsylvania, Iowa, and many other States I have received letters reporting cholera among turkeys, and often they say there are no symptoms of the trouble among the chickens. I have never believed it very kind in religious matters to try to shake one's faith in one thing without giving him something higher upon which to fasten his faith; yet in the matter of disease with turkeys I cheerfully try to convince my correspondents that it is not cholera which affects their flocks, for many facts appear to me to prove that it is not cholera, and yet I may not be able to tell exactly what it is. The fact that the turkeys linger for days, and sometimes even weeks, is one reason for believing it is not cholera. Another is that in many instances the chickens and turkeys are in the same yard and the chickens are not affected; then again, about the only symptom common to all inquiries is that the droppings are a yellowish green. Some describe the heads as black, saying they mope around and will not eat. Others say the head is red to the last, and they eat up to a few minutes before they are seized with an attack like convulsions; and still others say they have puffs under the eyes, while another flock has a white substance floating over the eye.

SYMPTOMS OF DISEASE.

It is a fact that almost any disease of a turkey will cause the droppings to become yellowish green, showing that disease in turkeys, like disease in the human family, sooner or later affects the digestive organs.

Often indigestion is the cause of the trouble. I am not positively certain that I ever had a genuine case of cholera in my yards, though I well remember when I thought every chicken or turkey that died had it.

I have been informed by a correspondent that there is a much larger per cent of deaths from what is known as blackhead than from cholera. He says what has often been pronounced cholera is blackhead. He also informs me there is absolutely no cure for it which can be relied on to be even comparatively a cure, and that the cause is unknown. This he wrote me some time since. He said that Lee's Germonoze is the best remedy known to him. I had some experience with the trouble in the flock of a neighbor, and I decided it was caused from over-feeding while young, and then turning them out without any food; at least, I found when I examined after death that the liver was perfectly soft and the gizzard twice the size it should have been.

USE OF RED PEPPER.

I find many persons use a great deal of red or cayenne pepper and soda in turkey food. Because I had been taught to do this, I did like my neighbors when I commenced raising turkeys, but I soon began, as Mr. Johnson suggests, to use my common sense, and I wondered how on earth anything could live, especially a wee bit of a turkey, with the crop filled with pepper, soda, sulphur, copperas, also custard, milk, curd, and many other things I was told I would have to feed to be successful. I said, "I shall try a way of my own," and whenever I have just had the sense to do my own way I have succeeded in keeping my flock healthy.

The greatest trouble I have is in early spring, when turkeys will eat dry grass. They become crop bound. I believe I have a sure cure for that, which I will give later.

GOOD MARKET FOR TURKEYS.

There will be more money go into farmers' hands from the sale of turkeys this year than for many years past. Turkeys are now selling on the market at 8 cents. Then talk about them eating their heads off! If a man sells hogs at 5 cents a pound he thinks he is doing a fine business, says he is getting 50 cents a bushel for his corn; but when a woman sells her turkeys at 8 cents a pound she is doing far better and is doing her husband a double favor; it is a favor to him for her to pay her own and part of his bills; besides, she gets 8 cents a pound for grasshoppers, which, but for her turkeys, would destroy the corn fodder and injure the hay, as well as ruin the cabbage.

This is fine growing weather for fowls, and turkeys especially. They can get all they need to eat by foraging now, and are making bone, and later they will take on flesh. Now is the time to select the oldest ones, and if they do not get plenty to eat, begin feeding them, for as soon as it turns cold there will be a cry for turkeys. People want a change in meat. They are tired of beef and of chickens and fish, and will pay a good price for turkeys. I expect to kill my first turkey for a family reunion on the 15th of October.

SELECTING THE SHOW BIRDS.

The fancier is looking over his or her flocks to see which ones are fit to show, and already inquiries are coming in asking, "What can you furnish me for the winter shows?"

It is a little too early to decide on the show birds, but still we think we can almost say which will score highest in our flocks. Persons often ask me if I will insure the turkeys I send to win first. No, I do not, and if others do they are either prophets are humbugs. I can only know what my stock is. I cannot know what it will meet in the show room, neither can I tell how a judge whom I have never seen score, nor seen any stock he has scored, will score.

It does seem to me, however, that this is wrong. If the Standard is a plain book and means what it says, it seems to me that it should be understood so that there can be no difference in scores of the same fowls in same condition and weight, and yet we know there is. The only guarantee one can give is that in his judgment it will score so much, and if it is beaten it will take a fine bird to do it.

I do not believe anything is gained by over-confidence. I do not hesitate to say that I believe I owe my success in a great measure to my feeling of doubt and desire to obtain only the best.

Well do I remember when I went to Sedalia with

three turkeys and five chickens. If any one had told me I would come back with a blue ribbon I would have thought he was making fun of me. Those three turkeys won four prizes and the chickens obtained one first.

This gave me the exhibition fever, which never goes down; not that I feel I have always had justice in the show room, for I have not, yet I can truthfully say I have never believed it to be the fault of the judge.

There is too much carelessness in weighing when so much depends on weight in show rooms. Give me just weights and I am willing to take chances under any judge I have known.

ALL BREEDERS SHOULD EXHIBIT.

Whether we win or whether we are beaten in show rooms, it is better to exhibit, and when I could attend a show I have never failed to sell enough to cover expenses of entrance fees. There we meet and exchange views with other fanciers, which is a great benefit to the one who is isolated from other fanciers at home.

It seems to me that the poultry business and especially turkey culture is one belt which reaches around the globe. America is sending fine turkeys to foreign countries, and the West is furnishing the East with her best turkeys. This year I shipped eggs and turkeys to New York, Pennsylvania, Kentucky, Virginia, Ohio, Indiana, California, Canada and Mexico, so that from Maine to Georgia (for I have shipped to each of these States), from the extreme West to the extreme East, turkeys are raised. They are one American bird which will thrive on any land, will grow on any soil, or in any climate, and are considered a luxury by all, from the king on his throne to the humblest laborer in his hut.

The practice by railroad officials and other large corporations of giving at Christmas one turkey to each man in their service, and where ladies are employed to them also, has lightened many a heart that otherwise would have felt the absence of the festal bird on the Christmas table. And this practice reverts with benefit to the farmer's wife, who with her turkey money can buy many Santa Claus presents which he otherwise could not distribute in her family.

The turkey is the king of American birds and America rules the poultry world, imported stock notwithstanding.

* * * * * * * *

Mr. Bancroft, of Memphis, Tenn., asks, "What must I do for our young turkeys and chicks? They turn blind, the flesh gets brown, they sit around a few hours and die. The tongue is pale and slimy. * * * We have lost two old turkeys the same way. * * * We use Lambert's Death to Lice and grease their heads with sulphur and lard. We feed them three times a day cornmeal and bran mash, with oats and chops mixed."

COOKED FOOD FOR POULTS.

If Mr. Bancroft gives the food to the little fellows uncooked it is not surprising that they die. Turkeys need very little bread, but if well baked and crumbled to them it will not hurt them. I consider bran injurious to turkeys and chicks unless it is baked with other ingredients.

Give the chicks the mixture above mentioned, prepared as follows: Put it in the oven dry and let it heat gradually until so hot that it will make a noise like anything frying on a stove when water is poured on it. Then pour just enough cold water on to thoroughly dampen it. Stir well and the heat will make a steam that will swell the bran and chops so that it will be fit for the crops of little chicks; it is also very good for poults if it is all you can get, but be careful not to overfeed. Little chicks and little turkeys must not be stuffed; let them scratch and get hungry. I feed my chicks only twice a day, when they have good range, for they are much more thrifty if not over-fed.

SULPHUR FOR LICE INJURIOUS.

I would not advise the use of sulphur in lard to grease the heads. The lard is sufficient and the sulphur takes away proper use of legs; it is also liable to cause blindness. Use clear lard or thick cream and grease the flight feathers just where they grow out from the wing bone. Insect powder is liable to cause blindness if it gets in the eyes. Use tobacco stems and tobacco dust as follows: Smoke the roosting-places with the stems, and sprinkle the floors with the dust.

Feed the little turkeys wheat bread soaked in milk or warm water. A hard-boiled egg mixed with it as a variety is very good. Cut up lettuce, onion tops or pepper grass with it. Plantain cut up fine is also good, but where little turkeys have a good place to run on the grass the green food is not so necessary. Give plenty of grit. I mix it in food or keep a supply near the roosting and feeding places.

SMALL EGGS MAY HATCH BIG TURKEYS.

I have been experimenting with some small turkey eggs. I found that some of my largest hens after they had laid several normal sized eggs laid very small ones —not as large as Brahma eggs. I had six of these on hand and would not ship them, but when I set the incubator I put them in it. When hatching time arrived I removed them and gave them to a chicken hen; she hatched six turkeys, and I find instead of them being very small weaklings they are good-sized, thrifty little poults, and if I can resist over-feeding, and keep them from rats, which seem determined to live with us, I think I will prove that a large turkey can be raised even though it was hatched from a very small egg.

By this time the hatching season for the fancier is over, both with chickens and turkeys. Of course, the hen may steal her nest and hatch a brood of chicks or poults, but the bulk of the hatching has been done and we are counting up the proceeds of the year's work in our minds.

Already we are planning how we shall provide winter quarters for chickens, and roosting sheds for turkeys. Many trials will beset us, do as we may. Only this morning I had a new experience with turkeys. Reading so much about fruit and poultry going so nicely together, I put my little turkeys in the meadow, which is set out in fruit trees. Seven cherry trees laden with large red cherries are admired by all passers-by, and I thought I had an ideal place for turkeys, where they would not be bothered by any of the other fowls. Yesterday I noticed a little turkey drooping. This morning I found it dead in the house. As it is my custom to try to find out the cause of the death of everything which dies on the place, I proceeded to hold an autopsy over the turkey. It had no vermin; was not thin in flesh; no signs of cholera or bowel trouble. On examination I found all the organs in healthy condition, but on cutting into the gizzard I discovered four cherry seeds. These could not pass out of the gizzard, and so killed the turkey. Now I am in a dilemma. I have no other suitable place to put these turkeys, and they are being taken care of by a chicken hen, which stays under the cherry trees, and I fear I shall lose all those little turkeys. This was one of six hatched from the very small eggs, and they have been growing so nicely I hoped to raise all of them. If I can keep them a week or ten days longer the cherries will be gone, and then I may rest contented. I am specially anxious to keep these, as I want to demonstrate that it does not take a large egg to produce a large turkey. I do not think the seed would hurt a turkey five or six weeks old, but these are not three weeks old, and the seed could not pass out of the gizzard.

LOW PRICES MEAN POOR STOCK.

If you send out fresh, fertile eggs, well packed, to honest customers, ninety-nine out of every hundred will either report a good hatch or tell you there were local reasons why they did not hatch. When people are willing to pay good prices for eggs, they are generally of that class who do not expect something for nothing. It is the 50-cent and dollar-a-sitting people who are hard to please, and it is the party who sells eggs so cheap who is liable to send out stale, infertile, poorly-packed ones.

This does not necessarily apply to the beginner, for I realize he has a reputation to make, and cannot command as high a price as one with a reputation made, but depend upon it that no matter how good your stock is, the fancier will not patronize you as long as you sell toms at $2.50 each. Why? Because he knows that you cannot afford to sell first-class birds at that price, and though you may have good stock he will be afraid to take the risk of buying from you.

Keep good stock, study the Standard, and be able to know whether yours is as good as it should be. Don't be a bigot, for the "know-all people" are a disgrace to any business or profession. Next to these the poultry business is weighed down with the "I-have-all-the-good-birds" people.

I may have different views from many, but I really believe that he who causes "two blades of grass to grow where only one grew before" is a benefactor to mankind, and so I believe that one who can without injury to himself start a fellow mortal in fine poultry is a benefactor, too.

A PREVENTIVE FOR FIGHTING TOMS.

A lady in Maryland writes: "I can't agree with you that hot mashes cause roup, for in my native home (Canada) I fed my turkeys hot mashes from the time they were a few weeks old until they were grown, and so did my neighbors, and we raised large flocks and were never troubled with roup. Here I have not fed the mash and have the roup." Roosting in trees may be the cause of roup in her flocks now, as she says they never had it when roosting under shelter. I agree with her that in extreme climates a shelter is better. She also gives me a preventive for fighting which I shall certainly use. It is as follows: Tie a small bell around the neck of the toms, showing a disposition to fight. The noise attracts their attention so they forget to fight.

A MYSTERIOUS DISEASE.

Another inquirer desires information concerning the disease affecting her turkeys, but I can only surmise the cause of death. She says: "They mope about for a few days and die. The discharge from bowels is first yellow, then green." She adds: "I do not believe it is cholera. The turkeys have been running on buckwheat fields, and some say this will kill turkeys," and asks if I have had any experience with buckwheat. I have not, and do not know whether it will kill turkeys, but have been under the impression that it would be good for them. Honey dew is said to be poisonous to them, and possibly this is the cause of death. Tadpoles will kill turkeys (so say old turkey raisers), and I am very much inclined to believe that many young turkeys die from insects or animal poisons of some kind which they get in orchard fields. I am not fully convinced that tadpoles are injurious, for the reason that my turkeys have always had access to ponds, and tadpoles are always abundant in stagnant water. I believe clear, running water better than the pond water for fowls, and I believe much of my success on the farm with turkeys has been due to running water; still the pond water was always in the fields in which they were.

THE BEST BREED.

Now comes a question, which I receive many times during the year, especially at this season: "Which do you consider the best breed of turkeys?" I am very frank to say that as I know comparatively little of any breed except the Bronze, I am not a competent judge of the best breed, but I do know that the Bronze is decidedly the most popular breed, and that which the majority like best must at least be the equal of any other breed. There are some objections made to the Bronze which may not be made to other breeds, and yet these objections are to the advantage of turkeys if properly managed. Many write, "I don't like the Bronze turkeys because they roam so much." And they certainly will roam if allowed free range, yet this roaming is very conducive to the vigor so much desired. Besides, the Bronze can be confined to small range with as much success as any other turkey. They can be trained to be as gentle and taught to stay at home if they are never allowed to visit, but if once allowed to get out on large range they are hard to make contented at home. My turkeys will eat out of my hands, and I can and often do bring flocks raised on farms to my place and in a few days they will be as gentle as those I raise. Turkeys are very susceptible to kind treatment and gentle management.

THE BEST BLOOD THE MOST PROFITABLE.

Another question asked is, "Do you advise me to get thoroughbred stock or grade turkeys?" I do not know how to advise another in this matter except to tell what I should do—I should never have "grade" anything on the place, except the highest grade of standard-breds I could get. Especially is this true of turkeys. I do not mean that I should under all circumstances raise only show birds in turkeys, but I should get the best blood and improve it every year if possible, and while I might not make show stock a specialty, yet they would be an even flock at no more expense than a common one and much more profit.

PROFITS IN POULTRY RAISING.

It is no longer a question, "Does poultry pay?" but the question now is, "How can I make it a better paying business than it now is?" True, we do see some advertising that they raise poultry for pleasure alone. When I arrive at that financial state where I am able to raise poultry for pleasure alone I do not intend to advertise it for sale, or have any anxiety over pleasing customers, lying awake at night, wondering if fowls shipped will meet the expectation of the customers and worrying when a customer writes, "I want a perfect turkey," after pricing a good breeding bird cheap. No, indeed! When pleasure alone is the object I shall raise my stock and let it go on the market or to those who ask for it. I shall not advertise for trade. While it is true that I derive pleasure from my present management, it is also true that there is not enough to do it for pleasure alone. There is too much anxiety, and there are too many men and women of many ways of doing business, for it to be all pleasure. True, the one in ten does stop to give a kindly word of thanks, and often the other side when in trouble remembers the kindnesses of other days and—asks for another favor! There is great pleasure in knowing that the labor of my hands has aided in making the family comfortable—still the idea given through advertisements of poultry for pleasure is that money is not a consideration with the advertiser, he or she doing it for the love of the poultry. Yet write to one of these for prices, and the very highest are given. So I conclude they, like myself, find the pleasure in the remuneration.

NINETY EGGS A SEASON.

A lady in Oklahoma writes, "I want to tell you that I have a turkey hen which I positively know laid ninety eggs this season." She wrote this at the suggestion I made in a former article of never having had those remarkably good layers. She gave me the pedigree of the turkey, and as it is pure Mackey Pride of the West stock I feel a little proud of that hen.

* * * * * * *

CONSTANT ATTENTION NECESSARY FOR SUCCESS.

I cannot, as I have often said, do much for a very little turkey, but I do not think where one has good strong parent stock that it is very hard to keep the little ones healthy. Not hard in the sense of an intricate work, but it is work continually—that is, no day must go by without cleaning the roosting house. I sweep the floor every morning, whitewash it once a week, and three or four times a week dust with insect powder or use a lice paint once a week, whichever is most convenient. If I can keep turkeys six weeks or two months I feel they will live, unless they have an accident. There must be no cessation of care from start to finish, and after they are large enough to go out into the wheat fields they are no trouble except to see that they are at home at night; yet one time neglecting them may cause the loss of the entire flock. Our farm is six miles distant, and I cannot have the benefit of the shattered wheat or that as I once did. My advice to all poultry raisers on large farms is to stay on them. Do not let any one induce you to think a small place is better. It has its advantages, but they are over-balanced by the disadvantages.

POST-MORTEM EXAMINATIONS.

I advise that persons losing turkeys hold a post-mortem. I do that with every one that dies unless I know the cause of death. If the parts show no disease I am sure that the balance of the flock will likely keep healthy and that death was due to some accidental cause rather than disease. If I find the parts diseased I begin to use prevention on the others. It is almost impossible to know the cause unless I do hold the post-mortem.

BEST METHOD OF MATING.

First, be sure your stock is strong and vigorous—never breed from a delicate turkey because it is well marked—it will not pay. If you are entering the fancy, size must be the first consideration. To be candid, I do not believe this the most sensible point to consider; but there is no use to butt my head against a stone wall when the only result will be to get my brains knocked out. So it is useless for one woman to say that it is not the wisest way to mate to have extremely large parent stock, when the Standard gives more points to weight than anything else in turkeys, and when every judge knows that the blue ribbon often goes to an inferior bird simply because he is up in weight. It has been several years since I selected my best marked turkeys for the show room. Why? Simply because I knew that more depends on weight than markings. Every fancier knows, too, that in spite of all we may write to the contrary, the largest turkeys are seldom the best marked, hence often the very best marked birds are killed on account of lack of weight.

DISTINCTION BETWEEN WEIGHT AND SIZE.

Again, it does not by any means follow that the heaviest turkey is the largest; but size is judged altogether by weight, and often the breeding stock is injured by overfeeding in order to advertise weight. In my opinion, fanciers are to blame for this demand for such heavy birds. They have created the demand for them by advertising them, and I do not hesitate to say, and my customers will bear me out in the statement, that

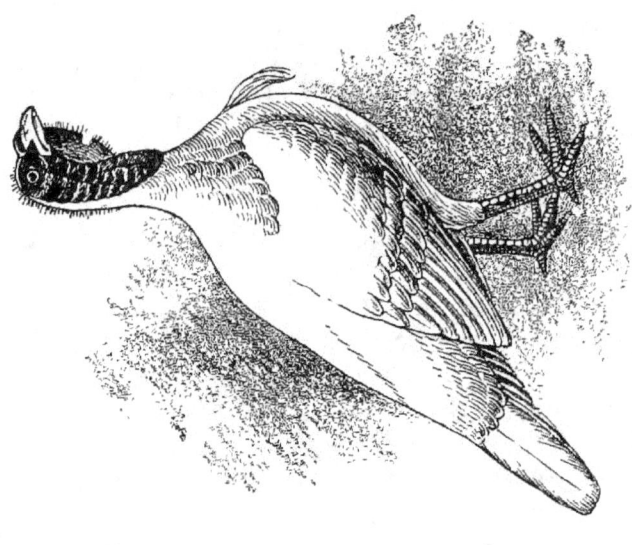
WHITE HOLLAND TURKEY.

BRONZE TURKEY.

in advertising these weights they often have but a few birds of weights claimed in the flock, and the customer is disappointed in the purchase. In climates where turkeys are hatched early they mature early and weigh in the fall and early winter what Western birds weigh in January and February; but we cannot hatch our turkeys so early, and consequently only a few of them attain maturity so early; but they are just as large-boned and vigorous, and are really better breeding stock for cold climates. Hence in saying I should make size the first consideration, I do not want to be understood as saying weight. After size I would make shape the next consideration, though size and shape go very nearly together. In both males and females select birds with broad breast and back. I prefer tall turkeys when young, for the tall, young turkey is the one that will make the heavy-weight bird when two years old.

SIZE, VIGOR, AND WEIGHT MUST NOT BE SACRIFICED TO FANCY MARKINGS.

Last come the markings. I find that Standard-marked females do not produce as finely-marked males as the darker ones, so in my flock of breeding birds I keep the females as nearly half and half with Standard-marked breasts and those darker than the Standard requires. To be more clear, the edging on the breast and back feathers of the female is required to be dull white or gray, and the wider the edging the higher the turkey scores. The Standard female produces finely-marked females, but the hen with a very narrow band of this edging produces the best-marked toms, the edging on the toms being black. Wing barring and tail markings are the hardest points to get. It is very hard to find both tail penciling and wing barring good on the same bird. One fancier attaches more importance to wing barring and another to tail penciling, and judges do not cut alike on these points, for judges differ widely in their interpretation of the Standard. I take the tom with the very best markings in both these sections that I can get, and mate him to the best-marked females I have.

I do not hesitate to use a female that would be disqualified in either wing or tail markings if she is extra large and otherwise well marked. If I am asked why I do this, I reply, because the tom will likely overcome the color disqualification (especially is this true of linebred stock where the blood is royal), and the female gives the size to young stock. Besides, if there are disqualified color markings and the bird has the size, there will be as many calls for that kind as for the best-marked ones, and if it must go on the market it will bring a good price. A fancier should never use any but a well-marked tom, for he is more apt to transmit color markings than the female; yet the female often transmits color markings.

SELLING CULLS.

Many fanciers say they will not sell culls at any price. By culls I mean birds not up in color markings. I sell birds that are not always up in fancy points, but I never sell one except to market breeders or persons who explicitly state that they do not care for fancy points. Again, it is well to state that I do not mean by culls birds which evidently are not pure bred stock, and this brings to mind the fact of stock having been exchanged on the road between the point of shipping and the place of destination. Two of the finest marked toms I sent out last season were proved to have been exchanged and birds put in their place which were not even thoroughbred Bronze. But the best bred Bronze may throw a turkey with wings with white splotches instead of evenly barred. These white splotches, while not a disqualification, are given as "very objectionable" in the Standard. I purchased last spring from a lady who had gotten her from a well-known fancier, a turkey hen which was sold to her as a high scoring bird. She had a straight gray feather in one wing. I have no doubt that she was thoroughbred, but it seems strange that a fancier would sell her for a high scoring bird. I wrote the lady about it and she replied, "I relied wholly on his statement concerning this hen and hope you will not think I misrepresented her to you."

SCORING BIRDS.

Last spring a lady wrote me that she wished to purchase a tom and wanted me to score it. I wrote her that I had never sent out a bird with a score card with my own name to it, but she insisted that I score it, as she wanted to put him on exhibition next year. I did so and am tremblingly waiting the result of the show so that I can know how nearly I came to the score of a professional. I should certainly prefer that the responsibility of scoring be on someone else. Even when Theo. Hewes and C. A. Emry sign score cards some people say they are of no account. One breeder said: "I would not trust a judge in America to select a tom for me, but I will describe what I want, and if you have it and will ship on approval I will take him." I declined. But while all fanciers are liable to mistakes, I believe that only a few are intentionally dishonest as compared with the number who advertise. And I can say that Missouri fanciers have always sent me just what they described, so have the Eastern fanciers from whom I have gotten stock, though sometimes I have been disappointed in results. But my motto is, "Try again."

A MYSTERIOUS MALADY.

There are many disappointments in all occupations, but at times there are more unexpected disasters in poultry breeding than almost any other business. Such was the feeling of several turkey breeders in this neighborhood a few weeks ago. Two flocks of turkeys (young ones) died without the least sign of disease, and upon examination after death no cause could be found. They were not lousy and seemed in perfect health; would eat heartily at night, refuse to eat in the morning and be dead before night. I confess I could do nothing for them, as I had no basis to work on. If anyone can help to solve the problem of the cause of death in these flocks at the age of from four to six weeks, the information will be gladly received.

I have received several inquiries from others describing death in their flocks as occurring in the same manner. The trouble was confined to turkeys, for the chicks on both farms were never in a healthier state and are growing very fast. Another flock of turkeys in perfect health was devoured by vermin. As I was directly interested in all of these flocks, I felt very much discouraged and was fearful that I would find the flocks on the other farms short; but a drive one day this week showed me that I need have no fear along this line and that larger and finer turkeys than I have ever had at this season of the year are now ready for shipping.

PREPARING FOR EXHIBITION.

This is a season of rest with the turkey raiser, if one is to be taken at all, for turkeys are now finding plenty to eat in the fields and have passed the danger of death from disease for a time at least; yet if any of them are to be put on exhibition very early they must be carefully tended by seeing that they have plenty to eat and are kept growing. They cannot be fattened while growing, but they may be kept growing by proper care. On large farms they can gather their own grain, and this year there are so many grasshoppers that they are well supplied with animal food.

SWELLED HEAD AND ITS REMEDIES.

Again and again the question comes, "What must I do for swelled heads in turkeys?" Some say the old birds have it; others, the young ones are attacked. I have given at different times as many as a dozen remedies, but will here repeat some of them. All are good, but so far I have found no remedy that cures all cases. I have used Conkey's Roup Cure, Littell's Liquid Sulphur and Mexican Mustang Liniment, and have effected cures with all of them and failed with all. I have used the salt water, tincture of iodine, sulphate of silver, with like results. I have just received some samples of Cushman's Roup Specific, and I believe from testimonials that it is all that is claimed for it; but as my fowls are now perfectly healthy I shall have to wait for an opportunity to try it.

FILTHY QUARTERS.

I do not like to go against public opinion, and I do

abominate filthy quarters for fowls; yet when I go through the country I find that some of the healthiest flocks I see do not have clean roosting places, and I conclude that chickens and turkeys are not half so tender as many fanciers suppose. Then I wonder how many fowls these people could raise if they took care of them.

THE TURKEY ALWAYS A WELCOME GUEST AT THE FEAST.

Can we be imaginative enough to picture to ourselves the difference between a family group at a Christmas dinner in 1799 and one at a Christmas dinner in 1899? Of one thing we are sure, and that is that the turkey was present at the former and also at the latter. We may go in imagination to the Christmas of 1999, and no doubt the turkey will be there. The turkey will be the main part of the feast at Thanksgiving and Christmas, at weddings and festal occasions, for time to come. We cannot recall a holiday time when it was not used, nor can we imagine one. Hence it certainly is not a matter of surprise that so much time should be given to the culture of turkeys.

Is it any wonder, then, that at this season letters of inquiry about breeding stock are coming thick and fast, and purchases are being made that the best results may be obtained? I think not.

SELECTING THE BREEDING STOCK.

I do not remember a time since I have had charge of this department when a month passed without some inquiry about disease with turkeys. Since my last writing, letter after letter has come to my table telling me of the healthy condition of turkeys and the success attained in raising them the past season. While this is very satisfactory in one sense, yet it leaves me without any pointers in the direction of my article, so far as health is concerned. But to be in line with inquiries received I shall answer some questions concerning breeding stock. In almost every letter I am asked to give description and weight of turkeys which I would recommend as breeding stock.

Right here I wish to say that I notice a practice among fanciers that may be all right, but I do not consider it the golden rule practice, to say the least. It is this, to send an inquiry on plain paper asking price of "good breeding birds," and after getting special price for such birds to send an order in a business envelope on a business letter head, with contents as follows or about as follows: "When sending these birds be sure that you send good ones, for I shall put them on exhibition as your birds." Now, it does seem to me that fanciers should not sail under false colors, and if they want exhibition birds should so state, for I am sure no breeder would not be willing to sell an exhibition bird for the price of a good breeding bird. More than once have I been caught in this net. Someone may ask, "Why did you not return the money?" For more than one reason I have not done this at times, and at other times I have returned the money. If I feel that it will be a good advertisement for me and I have stock enough on hand to send a better bird than priced, I would rather fill the order than lose a customer that way. In other cases I have sent just what I described and they were satisfied. But if one inquires for exhibition stock he should so state, for there are many excellent breeding birds (especially is this true of the Bronze turkeys) which are not exhibition birds.

FEWER CULLS AMONG TURKEYS THAN AMONG CHICKENS.

I hear some one say, "I think I have read that Mrs. Mackey says there were fewer turkey culls than chickens." And so she does; but while there are fewer culls among turkeys than there are among chickens, it is also true that under the present Standard some of the very best breeding birds are not first-class exhibition turkeys. For instance, all turkey raisers know that there is a tendency to brown edging on tail coverts in some specimens, and it is a notable fact that these specimens are usually very strong in wing barring and that they are decidedly larger boned than those which possess the Standard gray and white edging. Now such a bird cannot be sold for exhibition, but if one is raising turkeys for market and selling toms to other market breeders, what better breeder would he want than this same brown-tailed turkey? For as the market poultryman pays for pounds, it is the tom with size that one raising for market must get. Again, one may be poor in wing for exhibition and fine as a breeder. Yet I am sure that at least ninety per cent. of the best bred Bronze turkeys can be put on exhibition when fully up in weight, and whether they win a prize or not they do credit to the owner.

Again, I am asked how much a judge will cut for gray edging instead of white. Oh, me! How can I tell what a judge will do? On general principles, he will cut all he can. For my part, I do not see that he has a right to cut at all. The Standard says, "White or gray, white preferred." It seems to me if it allows gray as standard the judge has no right to cut for gray edging on tail. If only white is standard, say so in the Standard, and then we will know what is standard.

PINK LEGS.

Again, the question of pink legs comes up. Some write: "I want pullets and young toms with pink legs." The Standard says: "Shanks on young birds, dark, approaching black; in adult birds, usually a pinkish hue or flesh-color." I have had some pullets and young toms with shanks and toes almost as pink as the old birds'. I have some flocks that way last season, and I doubt not I shall find some this season. Customers at different times have written, "I never saw pink legs on young toms until this one." But the rule is, they are dark when young. Some of the best hens in size have dark shanks always, while others get very bright pink—I prefer the pink legs, other things being equal. But I am candid when I say that I make size and not weight the first consideration in breeding stock.

SELECTING FOR SIZE AND MARKINGS.

The tom for the fancier alone must be well marked,

as he impresses the markings on the young, and in selecting females I select size, then markings, and if I find an extra large bird of good shape, she does not go out of my breeding-pen because the edging is a little too dark on the tail feathers. And yet I find that every year the young stock comes nearer to Standard markings, there being a smaller per cent. each season that cannot be sent out as exhibition birds.

NEVER BREED FROM A DEFORMED BIRD.

There are certain rules I should advise every one to observe. Never breed from a turkey with a natural deformity. I once bought a sitting of eggs from one of the foremost fanciers I know. The only pullet raised had a crooked toe, but she was so fine I felt I could breed from her. Every year there would be a lot of turkeys with crooked toes, and these were the very best otherwise. It took several years to get entirely rid of crooked toes. It is such a temptation when one has paid a big price for eggs, and while I expect that under the same circumstances I might do it again, I believe it would be better not to use such a bird. The late Mrs. Foster told me that she would under such circumstances use the crooked-toed turkey as a breeder, as she could dispose of the deformed ones on the market and breed out the tendency to deformity with her own cross. I finally did this, but it took several years.

SHAPE FOR BREEDERS.

As to shape of tom and hens to breed from, I select large heads and feet, long body, long neck, held well up, and a broad back and breast, with long shanks. A short turkey will fatten earlier and look larger when not fully matured than the rangy one, but the latter will make the weight at maturity much heavier, and will produce larger turkeys. I select hens the same way; yet if they are specially well marked and good in weight I would not discard them if not quite as tall as I like them to be when pullets. I am sometimes disappointed in pullets, but cannot remember that I ever have been in a tom—pullets sometimes are no larger at two than at one year old. They often stop growing at one year, while a tom never does. Other pullets grow until they are two and three years old.

METHOD OF MATING.

My method of mating is simply this: Select the very best tom possible, and in females do not discard a very fine marked one because it is not quite as large as desired. By this I do not mean that I breed from small boned females. There are some larger than others in all flocks of the same age, and I should not advise the use in the breeding-pen of an undersized female, or a runt. Nor should I discard from my breeding-yard an extra large female because she is not quite up in fancy points, for the reason that the tom will overcome to some extent the defects. The well marked female will produce large stock from the mating with a large tom, and the one not so well marked will produce evenly marked young from the mating with a well marked tom. Yet these must be exceptional cases, for it will not do for a fancier to have many females in his breeding-yard that are not well marked and very large. By undersized, we mean pullets, for if at two years old a hen is not an average size I should discard her unless there were some special point I wanted to impress on my flock.

RED LEGGED PULLET.

I once had a red legged turkey pullet. She was not large when young, but her legs were almost too deeply colored to be called pink. I bred from her as long as she lived, or, I should say, until she was stolen. I could tell the turkeys from her eggs. They were a good size and invariably had pink legs when young, though not as deeply colored as were hers. From this hen I got that line of breeding which gives in some of my yards pink legs in young stock. I can tell it wherever I find it. But this is the only female I ever kept that was under size after she was a pullet. It is better if the breeding-yard can be made up entirely of extra large, well-marked birds, but so many persons ask me about mating that I have given these opinions.

THE GOLDEN RULE SHOULD GUIDE THE FANCIER.

Some ask me if I would use a late, small tom; another, a late, small pullet; and if I would use a poorly-marked pullet or a tom. These questions are asked by so many that I reply to all at once. Never should a fancier use a late, small-boned tom, or a poorly-marked one. Those raising for market only have nothing at stake except a few more dollars in or out of their personal accounts. The fancier should consider the interest of his customers and never breed from anything he or she would not be willing to have shipped to them. It is not so much a question of whether a customer is pleased as it is whether you have treated him as you would be willing to have him treat you were your positions reversed. We might please a customer simply because he would not know what he ought to receive, and one might be dissatisfied for the same reason. The fancier should know what he or she should ship, and do right because it is right.

SETTING TURKEYS ON THE GROUND.

One lady says: "Can you tell me what is the matter with my turkey hen? Her entrails are out for half a foot or more. I put them back, but they do not stay. Otherwise she seems perfectly well." No, I do not know what will cure her, because I never had any experience; but I think I should put her in a small place, where she could not exercise, and put the entrails back with vaseline or pure lard. If this did not cure her I should not know what to do. It may have been caused by her laying too large an egg. This same person asks if it will do to set a turkey on the ground if a nest is made. I prefer setting in a barrel, for if the weather is very wet there will be too much moisture for the eggs set on the ground. I have known the eggs to rot from dampness and have also known the poults to be so large they could not get out of the shell. Too much moisture will spoil the hatch out of the incubator as well as in it, but turkey eggs can stand much more moisture than chickens'. In fact, turkey eggs require more moisture than chicken eggs.

HATCHES FROM SMALL EGGS.

Another inquirer writes: "Will turkeys hatched from small eggs grow to be as large as those hatched from large eggs?" She says: "My breeding stock is large, the tom weighing thirty-five pounds and pullets and hens from sixteen to twenty-two pounds, but they lay small eggs. Do you know the cause? Do you think they will make large birds?" All I know in this line is from experience, and I have raised the largest turkeys I ever owned from small eggs.

A SATISFACTORY EXPERIMENT WITH SMALL EGGS.

Years ago I bought a setting of eggs from a prominent turkey fancier. I bought them in partnership with another lady. I was sorely disappointed when I opened them, and ashamed to let the other buyer see them, for I had influenced her in going in with me to make the order. So I told her I would release her and take all the eggs if she were not pleased, but she decided to take her share. We decided we were badly beaten, but as our money was gone we concluded to make the best of it by keeping still. I never saw such small turkey eggs. I made another order to another fancier, and never saw larger eggs than I received. The turkeys hatched from the small eggs grew to be larger at maturity, though, of course, they were not so large when hatched.

One tom from the small eggs weighed forty-four pounds at two years old and one hen twenty-four. Those from the large eggs never got so large, though they were fine turkeys. In speaking of this to an old lady and my surprise at results she replied: "I don't see why you should be surprised. Nature does not vary much, whether in lower animals, the feathered tribe or the human family. All depends on the blood." I believe the old lady was right. I prefer medium-sized eggs, both in turkeys and chickens. I find the very large eggs hatch large young, but they are not generally as well formed and often are weak-legged, and while medium-sized ones hatch smaller young they seem more vigorous and grow much faster. Of course, there are eggs that are small to deformity—so are there those that are large to deformity. Often the last egg laid before a hen goes to sitting will be so small it will have no yolk at all. I have had one that way this season. It is also true that the largest eggs are not always laid by the largest hens, though as a rule pullets lay smaller eggs than hens. In my yards this season I have found an exception to the rule, both with hen and pullet. One of the largest hens I have, that weighed twenty-four pounds in the winter, lays rather a small egg, and one of the pullets lays an unusually large one.

AN EXPERIMENT WITH A LARGE EGG.

I am now experimenting with the largest turkey egg I ever saw. It is larger than a goose egg. I am sure it must have two yolks, but I decided to try it, so set it. I am also experimenting with a broken egg; that is, a cracked one. I have up to this morning only those two turkey eggs being sat on at home. I found an egg cracked only slightly, so I pasted a thin cloth over the crack and put it with the large one under a sitting hen. I have hatched eggs that were cracked in an incubator, but my neighbors have insisted they will not hatch under hens. They also say the large egg will hatch a deformed turkey or not hatch at all. It is, I think, a universal complaint that turkeys in the North and West were unusually late in commencing to lay this spring, consequently the hatches will be later than usual. If we can have a good season and late fall those hatched in June will make show birds in January, provided proper care is taken of them, and July hatches make good breeding stock.

NUMBER OF EGGS TO A CLUTCH.

I am asked how many eggs a turkey hen will lay before she sits. Some lay more than others, and somehow I never get hold of any of those wonderful fowls I read about. I never had a hen that laid more than fifteen or sixteen eggs before she wanted to sit, and I have many more that lay only twelve than I have that lay fifteen. It may not be good policy to tell such a thing on my turkeys publicly, but I am telling facts. A turkey hen may easily be broken from sitting if she is taken in time, and by this means I often get twenty to thirty eggs before I allow a hen to sit. A hen will lay in ten days after she is stopped from sitting, and sometimes in shorter time.

SWOLLEN FOOT ON TOM.

Some one writes: "I have a fine young tom which has something the matter with one of his feet. The middle toe is swollen larger than the middle finger on a man's hand. I thought it might be bumblefoot. Please answer soon. State cause, disease and remedy." As this letter contained a stamped addressed envelope I should have been glad to reply if in my power. He says the tom has had the swelling several weeks. I do not know the cause, as I cannot tell without seeing the toe; but it is not bumblefoot, as that is on the bottom of the foot. If the tom were in my yard I should examine for a splinter or some foreign substance. If he is roosting high or even low he may have lighted on a nail, a splinter or something else and run it into his foot, causing the swelling. If I could not find anything of that sort I should examine for pus and cut the rising, letting the pus run out. I would then bind with Mexican Mustang Liniment, and I think with a few treatments he would be well.

DROOPING TURKEYS.

Still another inquirer says: "I wish you to tell me what to do for my young turkeys. Last year I lost nearly all of them. They did very well until about half grown and then they died. They ate heartily until they began to droop and then they did not eat any more until they died." I cannot tell from the facts given what was the matter with this lady's turkeys. I will just here tell my readers that when asking questions of one who cannot see your fowls, it is necessary you should know and describe the symptoms of the disease in such a manner as to make it plain to the person to whom you are writing. No one can say what is the matter with a fowl when she only knows that it quits eating and droops.

My way may not be the best, nor even only good, but it is the best I know, and I will give it for what it is worth.

When I have a drooping fowl I catch it and make a thorough examination. Sometimes it is a very small ailment and can be easily remedied. If after I have done what I could for it, it dies, then I have an autopsy and do my best to find out the cause of death. When this is discovered I think I can usually prevent other deaths from like causes.

SOFT-SHELLED EGGS.

I have not as yet begun to experiment with poults this season, but if my two eggs hatch I shall have some experience for next month; if they do not I have a neighbor who will have some poults in a short time and I shall have plenty of practice with them.

One more question is at hand: "Why do my hens lay soft-shelled eggs?" I suspect they are too fat, or perhaps they lack grit and lime. It will not do to give them lime unslacked or even air-slacked, but if you will put it in a vessel and pour water enough over it to make a thin mortar of it and let it set two or three days or longer, then give it to the fowls, you will see how eagerly they eat it, and it is very good for them. I dissolve in a candy bucket. When I put it out I put it on the ground in as near a solid lump as I can get it. The fowls soon have it broken up, but it falls in lumps and after it thoroughly dries it is quite hard; then I break the lumps into pieces the size of grit or larger. I feed little chicks on this place when I can, for the lime is good for them after it has been treated in this way.

SWELLING ON HOCK.

A breeder writes to know what to do for his turkeys. He says they have a swelling on the hock that seemed to have pus in it, but when opened it does not run.

I am entirely unacquainted with the trouble, and have no remedy except to bathe in Mustang Liniment.

My turkeys have been entirely free from disease this year, but pigs, cats and a varmint of some kind made havoc among them in the daytime for a period. I killed the cat, set the dog on the pigs, which action scared the varmint away, and now they are having a good time. I have taught the poults to come up without going to hunt them, and much strength and shoe leather are saved thereby. Turkeys can be taught to stay at home, and I have trained the hens that carry mine to come up at night, which is a comfort to me.

FEEDING AND CARE.

The turkey is by nature a wild animal, and is found in cold as well as in warm climates. It is an American bird, and I can remember when droves of wild turkeys were no uncommon sight in my native State, Virginia. These turkeys never had a hot mash nor a drink of hot water. They roosted in the highest pine trees they could find and drank from the flowing streams and springs in that mountainous region. They gathered the grain of the fields, and the insects which they caught served for meat. So if I were to select an ideal place for raising turkeys I should select a rocky, hilly place, with plenty of running water and plenty of grass, bounded by unlimited range, a place free from coyotes, foxes, minks, weasels and everything else which would destroy my flock. I should let them do just as they pleased, except I should feed and pet them just enough to keep them gentle.

I do not believe I should ever be troubled with cholera, roup or any of the diseases incident to tenderly raised fowls. But since this ideal place is not attainable I do the very best I can. I have demonstrated that thirty turkeys can be kept on less than an acre of ground during laying season by turning them out each day after they have laid in an orchard where they got water and ate apples and insects and plenty of grass. Out of this thirty not one was sick. Eleven were shipped the 1st of June, and the remaining nineteen are still healthy. However, they have had plenty of freedom since July, so that while I believe free range preferable, I have demonstrated that turkeys can be successfully handled in limited space. To do this they must be supplied with those articles of food which they get on free range. I have kept grit, oyster shells and lime before them constantly.

CAUSES OF ROUP.

For me to say that I do not know what causes roup and then tell you that I believe filth produces it may seem contradictory, yet I have noticed that when turkeys roost low in the same place and the droppings are either not removed or are kept covered with lime, those turkeys frequently are attacked with roup symptoms. I also observe that if turkeys are not taught to roost in the same place they frequently change their roosting places; hence I conclude it is better for them to do so, and we all know that the higher a turkey can get at night the better pleased he is, so I think pure air is very necessary for them. I think, too, that extreme changes in climate produce roup, yet I did not have a case of roup in my yards last winter, as cold as it was, and the care of my fowls for two months was left to my husband and little boy, who are not very much in love with poultry, and they thought when they gave them plenty of corn they had done their duty. Nothing but severe illness could have induced me to fail to give them my personal attention. The greatest loss I suffered was from toms fighting, and this caused swelled head, which finally killed them. By the way, I should like to ask if any one can give me a preventive for fighting

Mrs. B. G. Mackey.

Prevalence of Tapeworms in Turkeys.

I am convinced that tapeworms cause the death of great numbers of little turkeys, and that some suitable worm medicine should be frequently given them throughout the season. Turkeys are troubled with tapeworms from early spring until late in the fall, and sometimes have spasms from this cause. Very young turkeys suffer the most. After they are three months old they are better able to withstand the injurious effect. The worms apparently irritate the bowels, causing digestive derangement, diarrhea, weakness and death. At certain seasons segments of worms may be found early in the morning under the roosts among the droppings of the infected turkeys. Evidently the younger they receive the parasites, the more they suffer. Doubtless if the birds survive until the embryos have developed and have mostly passed out, they may gradually recover. A few worms may do little harm, while a great number may be fatal.

HOW CONTRACTED A QUESTION.

How the young receive the embryos in the spring is an interesting question. Whether snails, worms, or insects harbor them, and thus scatter the infection, or whether they receive the infection from the droppings of old turkeys, is yet to be determined. Keeping the turkeys on fresh, clean gravel, with light doses of freshly-powdered kousso or cusso as a medicine, is the best treatment.

REMEDIES FOR TAPEWORM.

Male fern is an effective remedy, but an overdose is a distinct poison. Six drams of the oil have caused the death of a person. It has been known to cause blindness in the lower animals and should be used with extreme caution. It is often given in combination with castor oil. Tansy is much used as a preventive and powdered areca nut for the removal of tapeworms from dogs and other animals. The latter is frequently combined with male fern. Ground pumpkin seed is also used as a remedy. The dose of these remedies

BLACK TURKEY.

WHITE HOLLAND TURKEY.

would have to be much reduced for turkeys. Turkey raisers may administer very light doses to a few turkeys and larger doses to others, and thus learn how great a quantity may be given to healthy turkeys with impunity. It is to be hoped that many may be able to apply these remedies with success and immediately prevent loss from this cause. Assafoetida, which is highly recommended for preventing and overcoming the gapeworm disease of fowls, is also said to possess virtues as a tapeworm remedy. This is administered either in the food or water. M. Megin, a French investigator, gave each pheasant seven and one-half grains of assafoetida combined with same quantity of pulverized gentian in their food, and overcame the gapeworm.

FOR LICE ON TURKEYS.

I am especially interested in turkeys, although we have three varieties of chickens, besides ducks. In all the articles I read concerning turkeys the writers say to look for lice on the top of the head and at the root of the flight feathers, but not one mentions the place where I find five to one to other places, and that is in the down on the young ones and at the root of the feathers on older ones on the large part of the thigh on the outside of the leg, about where the leg leaves the body.

The best food I have ever tried for the little poults is the curd that rises on the swill barrel. It is better than what is cooked on the stove. With this food and plenty of Lambert's Death to Lice on the hen before hatching time the little poults never have a thought of "turning up their toes." If you do not have Death to Lice use fresh lard every four or five days, applied where you find the lice.

Of course, dampness is another enemy to turkeys, but I have no experience in that line, as here in Montana we rarely have long continued storms.

F. E. CARVER.

Management.

Turkeys that are allowed to stray about too carelessly are very easily picked off by night prowlers.

The only way to raise first-class turkey stock is by giving poults plenty of exercise. The more of it they have, as a rule, the better they will grow and be in the end. Ranging about the fields, catching grasshoppers and other insects is the only thing that will give a growing poult the vigor that will develop it into a strong adult fowl.

RANGE FOR TURKEYS.

There is a general belief that exercise makes turkeys and other fowls tough, and therefore they should have no range to roam about in. This is true only to the extent that they become tough when taking an excessive amount of exercise. A range is necessary for the development of all fowls. The turkeys in particular need it. They should not be frightened, chased or worried while in the range, for that makes them tough and lean and spoils them as first-class birds. A flock of turkeys brought up naturally in a good range will weigh more, have finer and sweeter meat and be healthier all round than the birds brought up in the daintiest manner and spoon fed from their birth in narrow quarters. It is a natural life for them to roam about, and they should have all they want of it up to a period of a month before killing time. Those intended for the Thanksgiving market can be kept in the range until the 1st of November. They will improve by it and lay on flesh rapidly.—Wisconsin Agriculturist.

Turkeys in the Alfalfa Patch.

I fence a patch of alfalfa of four or five acres with wire netting, five feet high. If grass is not high enough to give good protection when turkey hens first begin to lay, I put a number of boxes around and let them lay under the boxes, and when they want to sit I fix a good nest right on the ground and leave the box over them to shade them, but leave box open on side at all times so hens can go to feed and water at will, unless two or more hens try to sit on same nest; in that event I pen one or two up in boxes and feed and water them in box until chicks are hatched.

By the time first sitting is off, the grass is high, and the hens hide their nests. I never disturb them, only to try and secure them when hatching, and put hen and chicks under a shade tree near ditch or pond, so they can take care of themselves. I usually carry about twenty or thirty hens and two or three gobblers in this pasture, and, in fair luck, I raise from 200 to 300 young turks every year.

Rearing Young Turkeys.

There is a general impression among farmers that turkeys are difficult to raise; so they are if the same method is adopted with them that holds good in raising chickens on the farm, but if the habits of the wild turkey are studied and the flock allowed to imitate them during the breeding season there will be but few obstacles to success.

The turkey has not overcome its wild nature to the same extent as the domestic fowl, and this fact must be considered. The domestic fowl when it hatches its brood of chicks is often cooped for a week or two, and when it is released it seldom wanders far from the coop, so that in case of storms or change of temperature shelter is easily reached. Not so with the turkey. If she is cooped for a time it makes little if any difference to her habits, and when she is released it seems that the further she can get away from her late habitation the better she is pleased. The poults have hitherto remained near the coop within call of the mother; they have been fed by the breeder, and the change of life which the wandering spirit of their parent renders necessary is entirely different from that they have been accustomed to, and calls for more robust constitutions than they possess. The mother has passed that period which cautions her to limit the exercise of her young, which she naturally does when the poults are first hatched, and instead of gradually increasing the length of her rambles in search for food, she at once forces her young beyond their strength, and runs chances of exposing them to weather to which they have not been accustomed. It would have been better to allow the mother her freedom from the first. The very weakness of the poults when hatched would keep them within reach of shelter for a time, and the weather hardening process would be gradual.

FREE-RANGE TURKEYS THE HARDIEST.

We have heard old breeders say that the morning dew and the dampness harbored by fields of hay and grain do not injure poults that have had their freedom from the start.

Turkeys with young will frequent the woods if there are any in the neighborhood, and among the trees they find sufficient vegetation to harbor insects while the ground is comparatively dry and free from long grasses. Such a location is therefore most desirable. Pasture fields, too, form a favorite range.

Half the battle of raising turkeys is won if the breeders are allowed free range for some time before the breeding season, and receive only sufficient food to induce them to return home in the evening. This treatment renders them hardy, removes the surplus fat that has accumulated during the heavy winter feeding, causes the eggs to be strongly fertilized and the poults to be on the jump from the time they are hatched. Rearing healthy poults is a pleasure, not a hardship.

If new corn is fed it should be given in small quantities until the turkeys have become quite thoroughly used to it.

To Have Big Turkeys.

Do not feed too much corn, but feed millet seed, oat groats to poults, with wheat and some cracked corn. Do not feed too much fattening food, but let them range out from the house. Do not feed with the chickens. Turkeys must have plenty of range to grow large and vigorous. Vigor in turkeys is the main thing to look after, and you will be sure to get size.—S. B. Johnston.

Bourbon Red Turkeys.

Mr. E. E. Page, of East Cleveland, who breeds Bourbon Red turkeys, has this to say about them:

"The history of the breed, as far as my knowledge of them runs, is as follows: About five years ago I moved to Knoxville, and while passing through Kentucky I noticed some dark red turkeys, and on making inquiry about them found that they were quite common in that neighborhood, and that they were a wild turkey which used to run all over the State. I became interested in them and contemplated putting them on the market, but since that time I have not had a location to take up the business. Lately I started to look up the turkeys again and wrote to several parties in Kentucky and found some splendid specimens. It had been some time since I had given any thought to them, and it was my impression that they were called Bourbon Reds, but on investigation I found that some called them Bourbon Butternuts and some Kentucky Reds. But I liked my name the best, and, having spoken of them several times under the name of Bourbon Reds, I decided to adopt that name. Since investigating the matter more thoroughly I find that they have been hunted in a region called Turkey Creek, in Southern Iowa; also in Southeastern Missouri and Northern Arkansas. Wherever they are known they are very much admired, both in the wild and domestic state, and truly they are a beautiful fowl, being a dark red color, the gobblers being the darkest and bordering on a brown, with white wings and tail, and the body feathers that have two narrow black bars running across the feathers, one on the tip being very black, the other close to it of a very light shade of black.

COLORING, SIZE, AND WEIGHT.

"The tips of all of the feathers have a bronze or bluish sheen, when shifted in the sunlight, similar to the bronze turkey. The under color runs out to nearly a white, with a majority of it a beautiful buff. In size, shape and weight when domesticated they resemble the good old bronze turkey, but are more hardy, better layers and less liable to wander away from home, and some claim that they stay as close to the house as they care to have them.

"The young are nearly as easy to raise as young chickens, and should meet with general favor on this account. This, together with their beautiful appearance and fine table qualities—large amount of breast meat—should make them sell well on the markets and to the fanciers.

"They are in no sense a mixed breed, but just as pure a variety as the bronze turkey, and no cross with bronze, buff or any other variety will produce a natural pure-blooded Bourbon Red turkey. They have been domesticated in Kentucky for eighteen years to my knowledge, and I am personally acquainted with a gentleman who saw them twenty-five years ago in Southern Iowa. The wild turkey is like the wild bronze—long, lean and lanky—and can outrun a deer, but the domestic Bourbon Red turkey is heavier breasted than the bronze turkey.

"Turkeys hatched in 1898 weigh now $36\frac{1}{4}$ pounds for gobbler and 18 pounds for hen; 1899 hatched gobblers, 25 pounds; hens, 14 pounds. I never saw anything that would outgrow them."—Poultry Keeper.

The Wild Turkey.

Inquiries in regard to the size and value of the wild turkey are numerous, and for the benefit of all interested will give our experience with them. We have been breeding for ten years, on different farms, a flock each of the pure bronze and wild, and for several years have been experimenting with the two crossed, which beyond any doubt makes the most beautiful, the largest, the most vigorous and magnificent bird of all the turkey family.

The pure wild turkey, when captured, is very hard to domesticate, and will oftentimes refuse to mate. But when hatched and raised from the eggs is quite easily domesticated and very handsome, indeed, being high in station, with the most perfect bronze plumage ever seen. In weight they are not so large as the bronze. The hen will average from 12 to 15 pounds at a year old, and the toms from 18 to 20 pounds. They are more hardy and vigorous than the bronze, and as a table bird are superb, having a delicious flavor. The great value of the wild is in introducing new blood or crossing with the bronze. But few realize what a magnificent bird this cross makes. The wild gives vigor, plumage and flavoring, while the bronze gives size and domestic qualities.—G. W. Brown in American Poultry Journal.

James H. Wilson, in "Fancy Fowls," says: "The only true bronze turkey is the wild turkey. No one ever heard of wild turkeys having cholera, roup or gapes. They are the healthiest, hardiest turkeys known, quick to detect danger and fight for their young. They are not hard to manage, as some people suppose. They don't stand around and wait to be fed. They get out to get their feed themselves and are easy to raise. By the time they are reduced one-fourth wild there is not much wild nature about them, although the color and constitution are greatly improved. I consider it advisable for turkey breeders to infuse wild blood in their flocks."

Habits of Wild Turkeys.

When a wild turkey hen is ready to lay she scratches out a slight hollow in a thicket—beside an old log—in tall grass or weeds or in a grain field, and pretends to line it with grass or leaves, and there proceeds to deposit her eggs, from ten to twenty in number, smaller but longer than those of our domestic turkey, but of the same color. Sometimes several hens lay their eggs in one nest and hatch and raise their broods together. Audubon found three hens sitting on forty-two eggs in a single nest, and one was always present to guard them. If the eggs are not destroyed, a single brood is raised in a year; but if they are, the female repeats her part of the performance, only being more exceedingly careful in hiding her nest and covering her eggs whenever she leaves the nest. Hunters "call" wild turkeys into range with peculiar whistles, and I have heard that this whistle, if successful, must be made of a turkey bone. They are also trapped. A turkey trap is nothing more nor less than a rail pen built strongly one rail square and as strongly covered, with the ground hollowed out under one side sufficiently deep to allow the turkeys to enter; then shelled corn is scattered not too plentifully outside in the ditch and inside. The hungry turkeys come and begin picking up the corn outside; then they settle down to work to get that in the ditch. By this time they are so busy eating they never raise their heads, but seeing the corn inside pass right under and into the pen. When their meal is finished we all know that turkeys raise their heads, and if frightened they never think of putting them down low, and thus it is they are easily secured. —Mrs. J. F. Knudeson, in Farmers' Advocate.

The F. P. C. PREPARATIONS,
For Chickens and Turkeys.
Manufactured by
F. P. CASSEL, Lansdale, Pa.

F. P. C. CHICK MANNA
10 DAYS FOOD.
For Little Chicks and Turkeys
when first hatched.
Promotes Health, Quick Growth,
and Strong Development.

1 lb. 10c., 5 lbs. 40c., 15 lbs. $1.10., 60 lbs. bulk per lb., 7c.

Office of "A Few Hens" Experimental Farm.
Hammonton, N. J., January 17, 1899.

F. P. CASSEL,
My Dear Sir:—I have used your Chick Manna all of last season with the very best results. The chicks not only like it, but they thrive wonderfully on it, and you can count on me being a steady customer. To get chicks rightly started is the secret of successful broiler and chick raising. The first two weeks is the critical period. We use the Manna exclusively for ten days and then gradually change over to other feed, and it gives us lively, strong chicks, which fully prepare themselves for the more forcing, fattening foods. Respectfully, MICHAEL K. BOYER,
Editor of "A Few Hens."

F. P. C. Agatha Poultry Food.
A specific for laying hens. Contains no oyster shell or ordinary mill feed. 5-lb. pkg. 35c. 60 lbs. bulk, per 1b. 6c.

I am using your Agatha Food with good results. I consider it a valuable preparation, and John says, "It is the thing to start the hens to lay." (Dec. 9, '99.) Very respectfully, DR. S. C. MOYER, Lansdale, Pa.
Breeder of Light Brahmas, W. and S. Wyandottes, and Barred and Buff Plymouth Rocks.

I have used your Agatha Food for some years and it is the best food I ever used to make the hens lay. My personal knowledge of the fact that it contains no ordinary mill feed or oyster shells gives me additional confidence in its value. P. M. FREDERICK, Lansdale, Pa.

F. P. C. Multum in Parvo Poultry Powder.
Valuable remedy for CHICKEN and TURKEY Cholera. Resorted to when other remedies fail. PURELY MEDICAL, no cheap material for bulk or weight in Multum in Parvo Powder. ½-lb. pkg. by mail, 28c.; 1 lb. by mail, 50c.

F. P. C. Roup Preparation.
A specific remedy for roup. Dissolved in drinking water, specially valuable for cankered mouth, and affected throat, valuable to use with Multum in Parvo Powder. ½-lb. pkg. 28c.; 1 lb. by mail, 55c.

F. P. C. Madoc Gape Cure.
Properly used produces QUICK EFFECT, SURE CURE. Full directions with each can. Try it. 6-oz. cans by mail, 28c.; ⅞-lb. cans by mail, 53c.

Prices do not include transportation charges, except where prices by mail are given. Descriptive circulars free on application.

Address communications to
F. P. CASSEL, Lansdale, Pa.

GRASSHOPPERS
ARE GREAT FOR
TURKEYS
BUT
Fidelity Food
IS GREATER.

Prominent and progressive turkey raisers report that they can rear poults far more successfully on Fidelity Food for Young Chicks than by any other system. Results—maximum of health and vigor, minimum of mortality. Fidelity Food for Young Chicks, used by leading fanciers and practical poultrymen, 25 lbs., $1.25; 50 lbs., $2; 100 lbs., $3.50. Fidelity Food for Fowls. Fidelity Fattening Food.

The famous Fidelity Foods are manufactured exclusively by the

Pineland Incubator & Brooder Co.,
JAMESBURG, N. J.

And are also kept in stock and for sale by leading poultry supply houses.

Mica-Crystal Co.,
CONCORD, N. H.
Manufacturers of
..MICA-CRYSTAL GRIT..
(Silica, Aluminum, Iron, and Magnesium.)

Standard Poultry Grit of America.

For Poultry, Pigeons, Chickens, Ducks, Geese, Turkeys, and Birds. In Four Sizes. Made from a rock that will not take a polish. THE ONLY GRIT containing the elements essential to the good health and egg-producing qualities of the feathered tribe. Send for descriptive circulars.

MICA=CRYSTAL CO.,
Concord, N. H.

If You Want the Best in White or Bronze Turkeys Give Me a Trial.

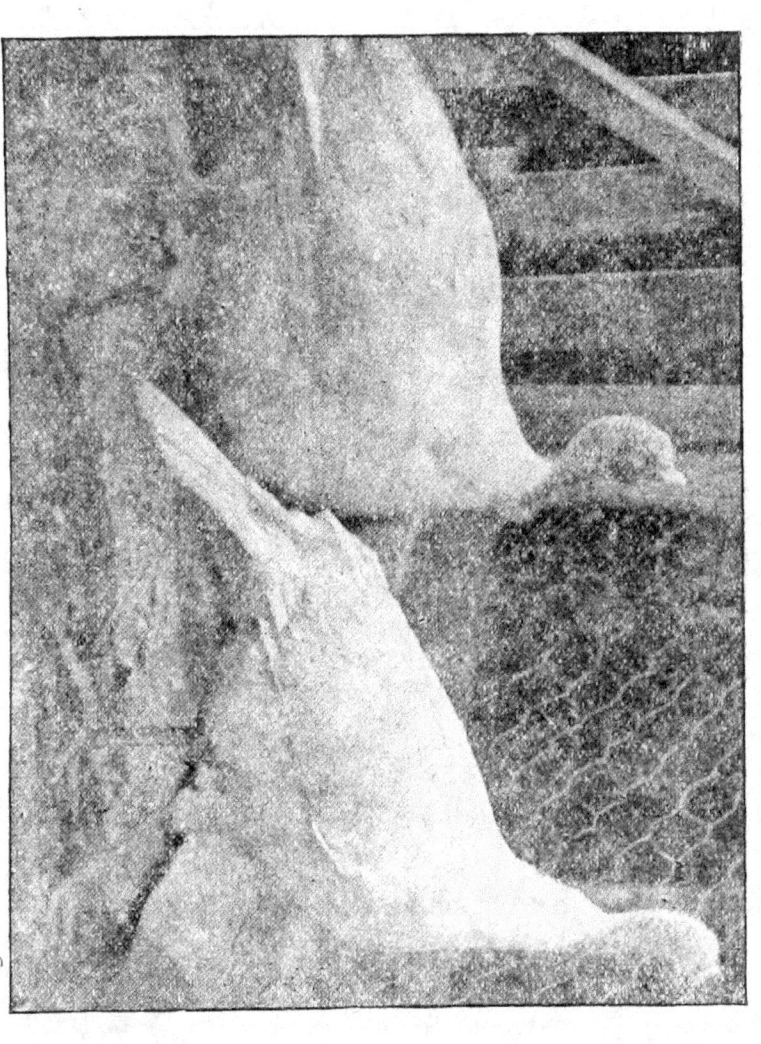

I have won more prizes with my White Turkeys than any other strain in this country. Birds and eggs for sale in season at reasonable prices. Also breed all the leading varieties of Land and Water Fowl. Send for large illustrated catalogue.

D. A. MOUNT,
Pine Tree Farm,

Lock Box 17,
JAMESBURG, N. J.

Wilson Bros.,
EASTON, PA., U. S. A.

Manufacturers of...

Grinding Mills...

For the

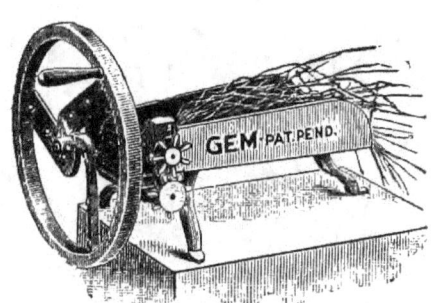

PRICE $9.00.
Weight 50 lbs.

Farmer,
House=keeper,
Poultryman,
Gardener,
Fertilizer
Manufacturer.

THE

Family Grist Mill.

Grinds fine corn meal and graham for table use...

The Poulter's Bone, Shell, Corn, and Grit Mill.

The Daisy Bone Cutter.

The Gem Clover Cutter.

Send for circular and testimonials.

MANN'S BONE CUTTERS

MAKE POULTRY...
BUSINESS A SUCCESS.

GREEN CUT BONE Doubles the Egg Product. Makes hens lay in winter when eggs are high. Makes early layers of pullets. Grows the quickest maturing broilers that will bring the highest prices.

MANN'S

Is the Oldest and Most Reliable Firm in the Business. Their machines are used from one end of the world to the other. They lead in all up-to-date improvements. They make a machine that runs easier and cuts faster than any other.

MANN'S MACHINE IS SENT ON TRIAL

in competition with any other or without competition. It never fails to please, and is never returned by a poultryman who really wants a machine. Send for free catalogue of Mann's Bone Cutters, Clover Cutters, Automatic Swinging Feed Trays, Granite Crystal Grit, Corn Sheller, etc.

F. W. MANN CO.,   Milford, Mass.

Green Bone
Makes Eggs.
The Dandy
Makes Cut Bone.
THERE ARE OTHERS....
But what a difference!

If the hired man runs it you may not appreciate it, but if you turn the crank get

THE DANDY GREEN BONE CUTTER.

It cuts rapidly with little power.
It cuts large bones direct from the butcher.
It has a perfectly automatic feed.
It has a tight-fitting receiving-pan, keeping all litter from the floor.
It is made of the very best material and fully warranted.
It has the endorsement of practical poultrymen throughout the country.
A trial will prove it to be a "Dandy."
Prices from $5 up to $100 according to size.
Illustrated catalogue free.

STRATTON M'F'G CO.,
ERIE, PENN.

MRS. MYRA V. NORYS, Westfield, N. J.

State Institute Lecturer and Authority

Poultry Books . . .

"Pocket=Money Poultry"
170 pages, 50c.

"Easy Poultry Keeping for Invalids,"
96 pages, 25c.

THESE BOOKS are terse, practical, helpful in an extraordinary degree; exactly what every beginner wants. They are bound in paper, and illustrated. Sent post paid at above prices. "Easy Poultry-Keeping" is published by the Author, who offers the usual discount to the trade. Special rates on large orders for use as premiums. Cash dealings only.

Standard-Bred...

White Wyandottes, Rose Comb Brown Leghorns.

Special attention given to furnishing exhibition birds, in both varieties. No eggs for sale. No breeds excel these in practical qualities. The **Rose Comb Brown Leghorns** are noted *the world over* as magnificent layers; the White Wyandottes both as remarkable layers and remarkably grand market birds.

MRS. MYRA V. NORYS,
Westfield, N. J.

THE $5 Champion Brooder.

"It is Known by Its Works."

THE STANDARD OF MERIT OF THE WORLD...

THE CHAMPION is the only brooder in the world which will successfully raise every chick whether the machine is placed out of doors in the severest winter weather, with the thermometer down to zero and below, or in your brooder-house. It is a perfect "zero weather" machine. There are more Champions in use than any other five makes. If you are in need of a brooder, send to the largest brooder manufacturers in the world for their elegant catalogue.

J. A. Bennett & Sons.
BOX IX, GOUVERNEUR, N. Y.

Spratts Patent Manufacture

Dog Cakes,
Charcoal Dog Cakes.
Greyhound Cakes,
Terrier Biscuit,
Plain Round Cakes,
Oatmeal Cakes,
Puppy Cakes,

Pet Do Cakes,
Cod Liver Oil Cakes,
Pepsinated Puppy Meal,
Orphan Puppy Food,
Plain Puppy Meal,
Crissel,
Cat Food,

Poultry Food,
Game Food,
Pigeon Food,
Pheasant Food,
Chick Meal,
Cardiac,
Bone Meal for Puppies.

Of Remedies, Etc., the Following:

Dog Soap (White),
Antiseptic Soap (Black),
Tonic Tablets,
Mange Cure,
Eczema Cure,
Purgative Tablets,
Cough Tablets,

Vermifuge,
Puppy Vermifuge,
Liniment for Sprains,
Cooling Tablets,
Locurium,
Hair Stimulant,
Jaundice Tablets,

Cure for Canker,
Fit Cure,
Distemper Tablets,
Anti-Rickets Tablets,
Eye Lotion Tablets,
Diarrhœa Cure.

A descriptive catalogue, containing a short treatise on dog diseases, feeding, etc., will be sent to any address on application.

SPRATTS PATENT American Ltd.,
450 to 456 Market Street, Newark, N. J.
Corner of Congress Street,

B. B. B. **CONTAINS EVERY PART OF AN EGG, ALBUMEN, YOLK AND SHELL.**

"What's the matter, children?" "We want some B. B. B."

GREATEST MEAT FOOD KNOWN for Laying Hens and Growing Chicks.   **B. B. B.**

BOILED BEEF AND BONE

Differs from all other similar poultry food, in that it is made from ABSOLUTELY FRESH MATERIAL, never over six hours old. The Cattle and Sheep Heads, Lights, Livers, and Beef are from stock slaughtered on the premises, and are cooked, dried, crushed, ground, mixed, and bagged, all within six to ten hours from time of killing. Samples sent free.

GUARANTEED cheaper than meat, better than scraps. Safer than medicine; rich in albumen. It prevents leg weakness, bowel complaint, feather eating, and assists in moulting.

50 lbs., $1.25; 100 lbs., $2.25.

D. W. ROMAINE,
Successor to SMITH & ROMAINE, Sole Manufacturers,
124 Warren St., New York City.

Banner Turkey Pills.

These Pills are made expressly for saving the lives of young turkeys. Turkeys, just after they are hatched, should be given a little stimulant, to give them a good start. Our Banner Turkey Pills contain just the right ingedients, having been carefully compounded, and are fully guaranteed to be the best life-saver for young poults. Directions with each box. These Pills are put up in two size boxes— **25c. and 50c. Boxes, postpaid.**

The Russ Chicken Cholera Cure.

The Russ Chicken Cholera Cure has demonstrated its superiority over all other remedies as a Cure and Preventive of Cholera, Gapes, Canker, Diphtheria, Diarrhœa, and Debility in fact, all bowel diseases that poultry are subject to. It is endorsed by the highest and best-known poultry and pigeon judges, fanciers, and authorities in the United States. When the Russ Chicken Cholera Cure is once used, none other is ever substituted. Experience has so well established its good qualities, that we do not deem it necessary to dwell upon its merits; all we ask is a fair and impartial trial, after which we know you will be a lifelong user of the Russ Chicken Cholera Cure. Every bottle guaranteed. **Price, 25c. per Bottle; cannot be sent by mail.** Our immense Illustrated Poultry Supply Catalogue. Send for one.

Excelsior Wire and Poultry Supply Co.,
26-28 Vesey Street,

W. V. RUSS, PROP.   NEW YORK CITY.

The Feather Library.

The following list of books are the most reliable publications of their kind in the market, and all our patrons should have them in their homes.

"DISEASES OF POULTRY."

By Dr. D. E. Salmon, Chief of the U. S. Bureau of Animal Industry. The only complete book on diseases and treatment. 248 pages; 72 illustrations. Price, cloth, $1; paper, 50 cents, postpaid.

"POCKET-MONEY POULTRY."

By Myra V. Norys. Tells how to make money raising poultry. The best book published, and is especially intended to assist women. 192 pages; 33 illustrations. Price, cloth, $1; paper, 50 cents, postpaid.

"THE AMERICAN FANCIER'S POULTRY BOOK."

By George E. Howard. A thorough and complete work on poultry raising. Every branch of the business explained. A book for the expert as well as novice. 170 pages; 215 illustrations. Price, 50 cents, postpaid.

"MONEY IN SQUABS."

By J. C. Long and G. H. Brinton. The only authentic treatise on raising squabs for market. Complete in every way. 93 pages; 32 illustrations. Price, 50 cents, postpaid.

"THE FEATHER'S ART PORTFOLIO."

This portfolio contains six beautiful pictures of poultry in half-tone and natural colors. Size 8x12 inches. Suitable for framing. Price each, containing six pictures, 50 cents, postpaid.

"THE AMERICAN STANDARD OF PERFECTION."

Latest revised edition. Published by the American Poultry Association. Every poultryman should have a copy of "The Standard" to breed his fowls successfully. Price, $1.

GEORGE E. HOWARD & CO.,
305 10th St. N. W., Washington, D. C.

www.ingramcontent.com/pod-product-compliance
Lightning Source LLC
Chambersburg PA
CBHW062329220526
45469CB00008B/2645